William Wallace Campbell

The elements of practical astronomy

William Wallace Campbell

**The elements of practical astronomy**

ISBN/EAN: 9783337278410

Printed in Europe, USA, Canada, Australia, Japan

Cover: Foto ©berggeist007 / pixelio.de

More available books at **www.hansebooks.com**

# THE ELEMENTS

OF

# PRACTICAL ASTRONOMY

BY

W. W. CAMPBELL

ASTRONOMER IN THE LICK OBSERVATORY

*SECOND EDITION, REVISED AND ENLARGED*

New York
THE MACMILLAN COMPANY
LONDON: MACMILLAN & CO., Ltd.
1899

*All rights reserved*

# PREFACE

## TO THE SECOND EDITION

My experience in presenting the elements of practical astronomy to rather large classes of students in the University of Michigan led me to the conclusion that the extensive treatises on the subject could not be used satisfactorily, except in special cases. Brief lecture notes were employed in preference. Arrangements were made with a local publisher that the notes should be written out in full and printed, almost exclusively it was supposed, for use in my own classes. The process of enlargement had just begun when the call to my present position was accepted. The completion of the manuscript in the midst of new and pressing duties was extremely difficult; the text and the details of the treatment lacked the harmony which can come only from a leisurely development of the subject. Nevertheless, the first edition has been used in a great many colleges and universities whose astronomical departments are of the highest character. This is my reason for carefully revising and slightly enlarging the book for a second edition.

A word concerning the limitations of the book may not be out of place. The field of practical astronomy has become very extensive, embracing essentially all the work carried on in our astronomical observatories. It includes the photographic charting of the stars, the spectroscopic determination of stellar motions, the determination of solar parallax from heliometer observations of the asteroids, the construction of empirical formulæ and tables for computing atmospheric refraction, and scores of other operations of equally high character. These, however, can best be described as special problems, requiring prolonged efforts on the part of professional astronomers; in fact, the solution of a single problem often severely taxes the combined resources of a number of leading observatories. While it is evident that a discussion of the methods employed in solving special problems must be looked for in special treatises and in the journals, yet these methods are all developed from the *elements* of astronomy, of physics, and of the other related sciences. It is intended that this book shall contain the elements of practical astronomy, with numerous applications to the problems first requiring solution.

It is believed that the methods of observing employed are illustrations of the best modern practice. The methods of reduction are intended to be exact to the extent that none of the value and precision of the observations will be sacrificed in the computations; further refinement would be superfluous, and misleading to the inexperienced. The demonstrations are direct and fundamental, except in the case of refraction. The scientific basis of the subject of refraction is largely physical, and the astronomical superstructure is almost wholly empirical. For these reasons, the proper proportions of the subject with reference to the rest of the book seem to be preserved by the insertion of the final formulæ.

An attempt has been made to give credit for methods which have not yet found their way into general practice. The illustrations of modern instruments are from cuts kindly furnished by the makers, viz.: those for Figures 15, 20, 21 and 24 by Fauth & Co., Washington; those for Figures 14, 17 and 26 by C. L. Berger & Sons, Boston; that for Figure 10 by the Keuffel & Esser Co., New York; and that for Figure 27 by Warner & Swasey, Cleveland. Figure 25, of a Repsold meridian circle, is copied with permission from Baron A. v. Schweiger-Lerchenfeld's *Atlas der Himmelskunde* (Vienna).

The author is indebted to his colleagues, Professors Schaeberle, Tucker and Hussey, and Mr. Perrine, for valuable suggestions and assistance.

W. W. CAMPBELL.

LICK OBSERVATORY,
UNIVERSITY OF CALIFORNIA,
January, 1899.

# CONTENTS

## CHAPTER I

|  | PAGE |
|---|---|
| THE CELESTIAL SPHERE | 1 |
| Definitions | 2 |
| Systems of coördinates | 7 |
| Transformation of coördinates | 8 |
| Distance between two stars | 14 |

## CHAPTER II

| TIME — Definitions | 15 |
|---|---|
| Conversion of time | 18 |

## CHAPTER III

| CORRECTION OF OBSERVATIONS — Form and dimensions of the earth | 23 |
|---|---|
| Parallax — Definitions | 25 |
|    In zenith distance | 26 |
|    In azimuth and zenith distance | 27 |
|    In right ascension and declination | 31 |
| Refraction | 32 |
|    General laws — In zenith distance | 33 |
|    In right ascension and declination | 36 |
| Dip of the horizon | 36 |
| Semidiameter — Of the moon | 38 |
|    Contraction of semidiameters by refraction | 39 |
| Aberration | 40 |
|    Diurnal aberration in hour angle and declination | 41 |
|    Diurnal aberration in azimuth and altitude | 43 |
| Sequence and degree of corrections | 43 |

## CHAPTER IV

PRECESSION — NUTATION — ANNUAL ABERRATION — PROPER
   MOTION . . . . . . . . . . 44
Star places . . . . . . . . . . . 45
Precession . . . . . . . . . 46
   Annual precession . . . . . . . . 50
Proper motion . . . . . . . . . 54
   Determination by the method of least squares . . 58
Reduction to apparent place . . . . . . . 61

## CHAPTER V

ANGLE AND TIME MEASUREMENT — The vernier . . . 65
The reading microscope . . . . . . . . 67
Eccentricity . . . . . . . . . . 69
The micrometer . . . . . . . . . . 70
   Determination of the value of a revolution . . . . 72
The level . . . . . . . . . . . 79
   Determination of the value of a division . . . . 81
The chronometer . . . . . . . . . . 83
Eye and ear method of observing . . . . . . . 85
The astronomical clock — The chronograph . . . . 86

## CHAPTER VI

THE SEXTANT — Description . . . . . . . 89
General principles of the sextant . . . . . . . 91
Methods of observing with the sextant . . . . . 92
Adjustments of the sextant . . . . . . . . 94
Corrections to sextant readings . . . . . . . 96
Determination of time . . . . . . . . . 101
   By equal altitudes of a fixed star . . . . . . 101
   By equal altitudes of the sun . . . . . . 102
   By a single altitude of a star . . . . . . . 106
   By a single altitude of the sun . . . . . . 107
Determination of geographical latitude . . . . 109
   By a meridian altitude of a star or the sun . . . . 109
   By an altitude of a star . . . . . . . . 110
   By circummeridian altitudes . . . . . . . 112
Determination of geographical longitude . . . . . 115
   By lunar distances . . . . . . . . . 115

CONTENTS xi

## CHAPTER VII

| | PAGE |
|---|---|
| THE TRANSIT INSTRUMENT — Description | 122 |
| Definition of instrumental constants | 127 |
| General equations | 129 |
| Determination of the wire intervals | 132 |
| Determination of the level constant | 134 |
| Determination of the collimation constant | 137 |
| Determination of the azimuth constant | 142 |
| Meridian mark, or mire | 143 |
| Adjustments | 144 |
| Determination of time | 146 |
| Reduction by the method of least squares | 152 |
| Correction for flexure | 157 |
| Personal equation | 157 |
| Determination of geographical longitude | 159 |
|     By transportation of chronometers | 159 |
|     By the electric telegraph | 160 |
|     By the heliotrope | 163 |
|     By moon culminations | 164 |

## CHAPTER VIII

| | |
|---|---|
| THE ZENITH TELESCOPE | 167 |
| Determination of geographical latitude by Talcott's method | 167 |
| Combination of results by the method of least squares | 173 |

## CHAPTER IX

| | |
|---|---|
| THE MERIDIAN CIRCLE — Description | 175 |
| Determination of right ascension | 177 |
| Determination of flexure | 181 |
| Errors of graduation | 183 |
| Determination of declination | 187 |

## CHAPTER X

| | |
|---|---|
| ASTRONOMICAL AZIMUTH | 191 |
| Azimuth by a circumpolar star near elongation | 193 |
| Azimuth by Polaris observed at any hour angle | 199 |

## CHAPTER XI

| | PAGE |
|---|---|
| THE SURVEYOR'S TRANSIT | 203 |
| Determination of time | 203 |
|     By equal altitudes of a star | 203 |
|     By a single altitude of a star | 205 |
|     By a single altitude of the sun | 207 |
| Determination of geographical latitude | 209 |
|     By a meridian altitude of a star | 209 |
|     By a meridian altitude of the sun | 210 |
| Determination of Azimuth | 211 |

## CHAPTER XII

| | |
|---|---|
| THE EQUATORIAL — Description | 212 |
| Adjustments | 214 |
| Magnifying power | 220 |
| Field of view | 221 |
| Determination of apparent place of an object | 223 |
|     By the method of micrometer transits | 223 |
|     By the method of direct micrometer measurement | 229 |
| Determination of position angle and distance | 233 |
| The ring micrometer | 236 |

## APPENDICES

| | |
|---|---|
| A. Hints on computing | 241 |
| B. Interpolation formulæ | 244 |
| C. Combination and comparison of observations | 247 |
| D. Objects for the telescope | 251 |
| TABLE I. Pulkowa refraction tables | 254 |
| TABLE II. Pulkowa mean refractions | 257 |
| TABLE III. Reduction to the meridian, or to elongation | 258 |

INDEX . . . . . . . . 261

# PRACTICAL ASTRONOMY

## CHAPTER I

DEFINITIONS—SYSTEMS OF COÖRDINATES—TRANS-
FORMATION OF COÖRDINATES

1. The heavenly bodies appear to us as if they were situated on the surface of a sphere of indefinitely great radius, whose center is at the point of observation. Their directions from us are constantly changing. They all appear to move from east to west at such a rate as to make one complete revolution in about twenty-four hours. This is due to the diurnal rotation of the earth. The sun appears to move eastward among the stars at such a rate as to make one revolution per year. This is caused by the annual revolution of the earth around the sun. The moon and the various planets have motions characteristic of the orbits which they describe. Measurements with instruments of precision enable us to detect other motions which, we shall see later, are conveniently divided into two classes: those due to parallax, refraction, and diurnal aberration, which depend upon the observer's geographical position; and those due to precession, nutation, annual aberration, and proper motion, which are independent of the observer's position.

From data furnished by systematic observations it has been shown that these motions occur in accordance with well-defined physical laws. It is therefore possible to compute the position of a celestial object for any given instant. A table giving at equal intervals of time the places of a body as affected by the second class of motions mentioned above, is called an **ephemeris** of the body. The astronomical annuals * furnish accurate ephemerides of the principal celestial objects several years in advance. If an observer knows his position on the earth, he can, from data furnished by the ephemerides, compute the direction of a star † at any instant. Conversely, by *observing* the directions of the stars with suitable instruments, he can determine the time and his geographical position. It is with this converse problem that we are principally concerned.

### DEFINITIONS

2. The sphere on whose surface the stars appear to be situated is called the **celestial sphere**. Any plane passing through the point of observation cuts the celestial sphere in a great circle. Since the radius is indefinitely great, all parallel planes whose distances apart are finite cut the sphere in the same great circle.

In order to determine the position of a point on the sphere and express the relation existing between two or more points, the circles, lines, points and terms defined below are in current use.

The **horizon** is the great circle of the sphere whose plane passes through the point of observation and is perpendicular to the plumb-line.

The produced plumb-line, or **vertical line**, cuts the sphere above in the **zenith** and below in the **nadir**. The

---

\* The principal annuals are the *American Ephemeris and Nautical Almanac*, the *Berliner Astronomisches Jahrbuch*, the (British) *Nautical Almanac*, and the *Connaissance des Temps*. Unless otherwise specified we shall refer to the first of these, and call it the **American Ephemeris**, or the **Ephemeris**.

† For convenience we shall use *star*, *point* or *body* to denote any celestial object.

zenith and nadir are the poles of the horizon, and all great circles passing through them are called **vertical circles**.

The points of the horizon directly south, west, north and east of the observer are called, respectively, the **south, west, north** and **east points**.

The **meridian** is the vertical circle which passes through the south and north points.

The **prime vertical** is the vertical circle which passes through the east and west points.

The **altitude** of a point is its distance from the horizon, measured on the vertical circle passing through the point. Distances *above* the horizon are $+$; *below*, $-$. The altitudes of all points on the sphere are included between $0°$ and $+90°$, and $0°$ and $-90°$. Instead of the altitude, it is frequently convenient to use the **zenith distance**, which is the distance of the point from the zenith, measured on the vertical circle of the point. It is the complement of the altitude. The zenith distances of all points on the sphere lie between $0°$ and $+180°$.

The **azimuth** of a point is the arc of the horizon intercepted between the vertical circle of the point and some fixed point assumed as origin. With astronomers it is customary to reckon azimuth from the south point around to the west through $360°$. Surveyors frequently reckon from the north point.

The **celestial equator** is the great circle of the sphere whose plane is perpendicular to the earth's axis. It therefore coincides with or is parallel to the terrestrial equator.

The earth's axis produced is the **axis of the celestial sphere**. It cuts the sphere in the north and south poles of the equator. We shall for brevity call them the **north** and **south poles**.

All great circles passing through the north and south poles are called **hour circles**. The hour circle passing through the zenith coincides with the meridian.

The **declination** of a point is its distance from the equator, measured on the hour circle passing through the point. Distances *north* are $+$; *south*, $-$. The declinations of all points on the sphere are included between $0°$ and $+90°$, and $0°$ and $-90°$.

Instead of the declination, it is sometimes convenient to use the **north polar distance,** which is the distance of a point from the north pole, measured on the hour circle of the point. It is therefore the complement of the declination. The north polar distances of all points lie between $0°$ and $+180°$.

The **hour angle** of a point is the arc of the equator intercepted between the meridian, or south point of the equator, and the hour circle passing through the point. In practice, however, it is customary to consider the hour angle as the equivalent angle at the north pole between the meridian and hour circle. It is reckoned from the meridian around to the west through 24 hours, or $360°$.

The **ecliptic** is the great circle of the sphere formed by the plane of the earth's orbit; or, it is the great circle described by the apparent annual motion of the sun. It intersects the equator in two points called the **equinoxes.**

The **vernal equinox** is that point through which the sun appears to pass in going from the south to the north side of the equator (about March 20).

The **autumnal equinox** is that point through which the sun appears to pass in going from the north to the south side of the equator (about Sept. 22).

The **solstices** are the points of the ecliptic $90°$ from the equinoxes. The sun is in the **summer solstice** about June 21; in the **winter solstice** about Dec. 21.

The **equinoctial colure** is the hour circle passing through the equinoxes. The **solstitial colure** is the hour circle passing through the solstices.

The angle between the equator and ecliptic is called the **obliquity of the ecliptic.**

The **right ascension** of a point is the arc of the celestial equator intercepted between the vernal equinox and the hour circle of the point. It is measured from the vernal equinox toward the east through 24 hours, or 360°.

The **sidereal time** at any point of observation is equal to the right ascension of the observer's meridian. It is likewise equal to the hour angle of the vernal equinox.

Great circles perpendicular to the ecliptic are called **latitude circles**.

The **latitude** of a point is its distance from the ecliptic, measured on the latitude circle passing through the point. Distances *north* are $+$; *south*, $-$. The latitudes of all points on the sphere are included between 0° and $+90°$, and 0° and $-90°$.

The **longitude** of a point is the arc of the ecliptic intercepted between the vernal equinox and the latitude circle of the point. It is measured from the vernal equinox toward the east through 360°.

The position of an observer on the earth's surface is defined by his geographical latitude and longitude.

The **geographical latitude** of a place is the declination of the zenith of the place. It is also equal to the altitude of the north pole. Latitudes of places *north* of the equator are $+$; *south*, $-$.

The **geographical longitude** of a place is the arc of the equator intercepted between the meridian of the place and the meridian of some other place assumed as origin. It is customary to reckon longitudes *west* $(+)$ and *east* $(-)$ from the meridian of Greenwich, through 12 hours, or 180°.

The preceding definitions are illustrated by Fig. 1. The celestial sphere is orthogonally projected on the plane of the horizon, $SWNE$. The zenith $Z$ is projected on the point of observation. $NZS$ is the meridian; $EZW$ the prime vertical; $WVQE$ the equator; $VLBV'$ the ecliptic; $P$ the north pole; $P'$ the north pole of the ecliptic; $V$ the vernal equinox; $V'$ the autumnal equinox; $VP$ the equinoctial colure; $CPP'$ the solstitial colure; $BC = PP' = BVC\ast =$ the obliquity of the ecliptic.

Let $O$ be any point on the sphere; then $ZOA$ is its vertical circle; $MOP$ its hour circle; $LOP'$ its latitude circle. The position of the point $O$ is defined by the following arcs, called **spherical coördinates**:

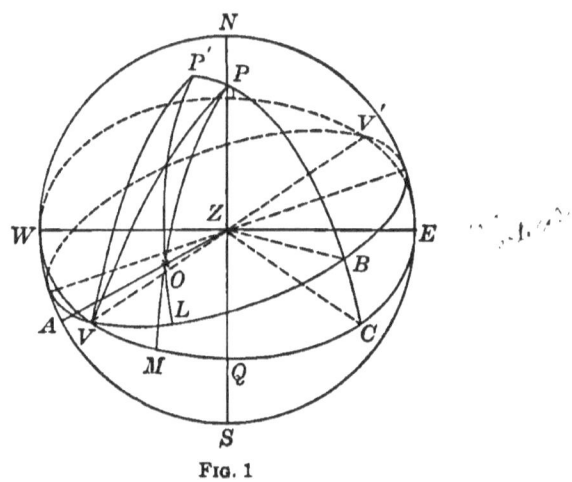

Fig. 1

$$AO = \text{Altitude, } h,$$
$$ZO = \text{Zenith distance, } z,$$
$$SA = SZA = \text{Azimuth, } A,$$
$$MO = \text{Declination, } \delta,$$
$$PO = \text{North polar distance, } P,$$
$$QM = QPM = \text{Hour angle, } t,$$
$$VM = \text{Right ascension, } a,$$
$$VQ = VPQ = \text{Sidereal time, } \theta,$$
$$LO = \text{Latitude, } \beta,$$
$$VL = \text{Longitude, } \lambda,$$
$$PP' = BVC = \text{Obliquity of the ecliptic, } \epsilon.$$

---

\* To be exact, we should say that the angle $BVC$ *is measured by* the arc $BC$ and by the arc $PP'$; but in the operations of practical astronomy the distinction is seldom made.

SYSTEMS OF COÖRDINATES 7

**3.** It will be observed that the horizon, equator and ecliptic are of fundamental importance. They are called **primary circles**. Vertical circles, hour circles and latitude circles, which are respectively perpendicular to them, are called **secondary circles**. Two spherical coördinates, one measured on a primary circle, the other on its secondary, are necessary and sufficient to determine completely the direction of a point; and from the definitions just given, to meet the requirements of astronomical work, we formulate four

SYSTEMS OF COÖRDINATES

| | Circles of Reference | | Coördinates | |
|---|---|---|---|---|
| System | Primary | Secondary | Primary | Secondary |
| I | Horizon | Vertical circle | Azimuth | Altitude |
| II | Equator | Hour circle | Hour angle | Declination |
| III | Equator | Hour circle | Right ascension | Declination |
| IV | Ecliptic | Latitude circle | Longitude | Latitude |

The altitude, azimuth and hour angle of a star are continually changing. They are functions of the time and the observer's position. Hence they are adapted to the determinations of time, azimuth and geographical latitude and longitude. Right ascension and declination are nearly independent of the observer's position, and vary with the time. They are largely used for recording the relative positions of stars, and in ephemerides. Latitude and longitude are also nearly independent of the observer's position, but are employed almost exclusively in theoretical astronomy.

In the solution of many problems of practical astronomy, it is required that the coördinates of a point in one system be transformed into the corresponding coördinates in another system.

## TRANSFORMATION OF COÖRDINATES

**4.** *Given the altitude and azimuth of a star, required its declination and hour angle.*

This transformation is effected by solving the spherical triangle $PZO$, Fig. 1, whose vertices are at the pole, the star, and the zenith. Three parts of this triangle are known: $ZO$ the zenith distance or complement of the given altitude, $PZO$ the supplement of the given azimuth, and $PZ$ the complement of the given latitude; from which, by the methods of Spherical Trigonometry, we can find $PO$ the complement of the required declination, and $ZPO$ the required hour angle.

For any spherical triangle $ABC$ we have [*Chauvenet's Sph. Trig.*, § 114] the general equations

$$\cos a = \cos b \cos c + \sin b \sin c \cos A, \qquad (1)$$
$$\sin a \cos B = \cos b \sin c - \sin b \cos c \cos A, \qquad (2)$$
$$\sin a \sin B = \sin b \sin A. \qquad (3)$$

To adapt these equations to the triangle $POZ$, let

$$A = PZO = 180° - A, \qquad a = PO = 90° - \delta,$$
$$b = ZO = 90° - h, \qquad B = ZPO = t,$$
$$c = PZ = 90° - \phi.$$

Then (1), (2) and (3) become

$$\sin \delta = \sin h \sin \phi - \cos h \cos \phi \cos A, \qquad (4)$$
$$\cos \delta \cos t = \sin h \cos \phi + \cos h \sin \phi \cos A, \qquad (5)$$
$$\cos \delta \sin t = \cos h \sin A, \qquad (6)$$

which enable us to find $\delta$ and $t$.

If $h$ be replaced by its equivalent, $90° - z$, these become

$$\sin \delta = \cos z \sin \phi - \sin z \cos \phi \cos A, \qquad (7)$$
$$\cos \delta \cos t = \cos z \cos \phi + \sin z \sin \phi \cos A, \qquad (8)$$
$$\cos \delta \sin t = \sin z \sin A. \qquad (9)$$

These equations are not adapted to logarithmic computations (unless addition and subtraction logarithmic tables are employed), and they will be further transformed.

Let $m$ be a positive abstract quantity, and $M$ an angle such that

$$m \sin M = \sin z \cos A, \qquad (10)$$
$$m \cos M = \cos z, \qquad (11)$$

which conditions may always be satisfied [*Chauvenet's Plane Trig.*, § 174]. Substituting these in (7), (8) and (9), they become

$$\sin \delta = m \sin (\phi - M),$$
$$\cos \delta \cos t = m \cos (\phi - M),$$
$$\cos \delta \sin t = \sin z \sin A.$$

From these and (10) and (11) there result

$$\tan M = \tan z \cos A, \qquad (12)$$
$$\tan t = \frac{\tan A \sin M}{\cos (\phi - M)}, \qquad (13)$$
$$\tan \delta = \tan (\phi - M) \cos t, \qquad (14)$$

which completely effect the transformation. The computations are partially checked by (9).

The quadrant of $M$ is determined by (10) and (11). $t$ is greater or less than 180° according as $A$ is greater or less than 180°, since both terminate on the same side of the meridian. The quadrant of $\delta$ is fixed by (14).

*Example.* At Ann Arbor, 1891 March 13 the altitude of *Regulus* is + 32° 10′ 15″.4, and the azimuth is 283° 5′ 6″.4. Find the declination and hour angle. [For instructions in the art of computing, see Appendix A.]

| | | | | |
|---|---|---|---|---|
| $\phi$ + | 42° 16′ 48″.0 | (Amer. Ephem., p. 482) | | |
| $z$ | 57 49 44 .6 | | $\tan (\phi - M)$ | 9.616914 |
| $A$ | 283 5 6 .4 | | $\cos t$ | 9.728805 |
| $\tan z$ | 0.201331 | | $\delta$ | +12° 29′ 56″.4 |
| $\cos A$ | 9.354873 | | | |
| $M$ | 19° 47′ 40″.1 | | | Proof |
| $\phi$ | 42 16 48 .0 | | $\sin z$ | 9.927608 |
| $\tan A$ | 0.633702$_n$ | | $\sin A$ | 9.988575$_n$ |
| $\sin M$ | 9.529753 | | $\csc t$ | 0.073401$_n$ |
| $\sec (\phi - m)$ | 0.034339 | | $\sec \delta$ | 0.010417 |
| $\tan t$ | 0.197794$_n$ | | $\log 1$ | 0.000001 |
| $t$ | 302° 22′ 54″.0 | | | |
| $t$ | 20$^h$ 9$^m$ 31$^s$.6 | | | |

**5. Given the declination and hour angle of a star, required its azimuth and zenith distance.**

In the general equations (1), (2) and (3) let

$$b = 90° - \delta, \quad c = 90° - \phi, \quad A = t,$$
$$B = 180° - A, \quad a = z;$$

and they become

$$\cos z = \sin\delta \sin\phi + \cos\delta \cos\phi \cos t, \tag{15}$$
$$\sin z \cos A = -\sin\delta \cos\phi + \cos\delta \sin\phi \cos t, \tag{16}$$
$$\sin z \sin A = \cos\delta \sin t. \tag{17}$$

To transform them for logarithmic computation, put

$$n \sin N = \sin\delta, \tag{18}$$
$$n \cos N = \cos\delta \cos t. \tag{19}$$

Whence

$$\tan N = \frac{\tan\delta}{\cos t}, \tag{20}$$

$$\tan A = \frac{\tan t \cos N}{\sin(\phi - N)}, \tag{21}$$

$$\tan z = \frac{\tan(\phi - N)}{\cos A}, \tag{22}$$

which effect the transformation. (17) furnishes a partial check on the computations.

*Example.* At Ann Arbor, 1891 March 13, when the hour angle of *Regulus* is $20^h 9^m 31^s.6$, what are the azimuth and zenith distance?

$$\delta \quad + 12° 29' 56''.4 \text{ (Amer. Ephem., p. 332)}$$
$$t \quad 302\ 22\ 54\ .0$$

| | | | |
|---|---|---|---|
| $\tan\delta$ | 9.345719 | $\tan(\phi - N)$ | 9.556204 |
| $\cos t$ | 9.728805 | $\cos A$ | 9.354873 |
| $N$ | 22° 29′ 7″.9 | $z$ | 57° 49′ 44″.6 |
| $\phi$ | 42 16 48 .0 | | Proof |
| $\tan t$ | $0.197794_n$ | $\cos\delta$ | 9.989583 |
| $\cos N$ | 9.965661 | $\sin t$ | $9.926599_n$ |
| $\operatorname{cosec}(\phi - N)$ | 0.470247 | $\operatorname{cosec} A$ | $0.011425_n$ |
| $\tan A$ | $0.633702_n$ | $\operatorname{cosec} z$ | 0.072392 |
| $A$ | 283° 5′ 6″.4 | $\log 1$ | 9.999999 |

## TRANSFORMATION OF COÖRDINATES

**6.** The angle $POZ$, Fig. 1, between the hour and vertical circles of a star, is called the star's **parallactic angle**. Let $q$ represent it.

To find the parallactic angle when $z$, $A$, and $\phi$ are given, we have, from (1), (2) and (3),

$$\sin \delta = \cos z \sin \phi - \sin z \cos \phi \cos A, \qquad (23)$$

$$\cos \delta \cos q = \sin z \sin \phi + \cos z \cos \phi \cos A, \qquad (24)$$

$$\cos \delta \sin q = \sin A \cos \phi. \qquad (25)$$

Assume

$$k \sin K = \sin \phi, \qquad (26)$$

$$k \cos K = \cos \phi \cos A, \qquad (27)$$

and we obtain

$$\tan K = \frac{\tan \phi}{\cos A}, \qquad (28)$$

$$\tan q = \frac{\tan A \cos K}{\cos (K - z)}. \qquad (29)$$

The quadrant of $q$ is determined by (25) and (29).

To find the parallactic angle and zenith distance when $\delta$, $t$, and $\phi$ are given, we have, from (1), (2) and (3),

$$\cos z = \sin \delta \sin \phi + \cos \delta \cos \phi \cos t, \qquad (30)$$

$$\sin z \cos q = \cos \delta \sin \phi - \sin \delta \cos \phi \cos t, \qquad (31)$$

$$\sin z \sin q = \sin t \cos \phi. \qquad (32)$$

Assume

$$l \sin L = \cos \phi \cos t, \qquad (33)$$

$$l \cos L = \sin \phi, \qquad (34)$$

and we obtain

$$\tan L = \cot \phi \cos t, \qquad (35)$$

$$\tan q = \frac{\tan t \sin L}{\cos (\delta + L)}, \qquad (36)$$

$$\tan z = \frac{\cot (\delta + L)}{\cos q}. \qquad (37)$$

The computations may be partially checked by (32).

The values of $q$ obtained from the data of § 4 and § 5 are equal to each other, and to $312° \ 25' \ 33''.5$.

**7.** *Given the declination and zenith distance of a star, required its hour angle.*

If a, b and c are the sides and A an angle of a spherical triangle, we have [*Chauvenet's Sph. Trig.*, § 18]

$$\tan \tfrac{1}{2} A = \pm \sqrt{\frac{\sin(s-b)\sin(s-c)}{\sin s \sin(s-a)}},$$

in which $s = \tfrac{1}{2}(a+b+c)$. If in this we substitute from triangle $POZ$

$$A = t, \quad a = z, \quad b = 90° - \delta, \quad c = 90° - \phi,$$

it reduces to

$$\tan \tfrac{1}{2} t = \pm \sqrt{\frac{\sin \tfrac{1}{2}[z+(\phi-\delta)]\sin \tfrac{1}{2}[z-(\phi-\delta)]}{\cos \tfrac{1}{2}[z+(\phi+\delta)]\cos \tfrac{1}{2}[z-(\phi+\delta)]}}. \qquad (38)$$

Similarly, it can be shown that

$$\sin \tfrac{1}{2} t = \pm \sqrt{\frac{\sin \tfrac{1}{2}[z+(\phi-\delta)]\sin \tfrac{1}{2}[z-(\phi-\delta)]}{\cos \phi \cos \delta}}. \qquad (39)$$

To determine the quadrant of $t$ it must appear from the data of the problem whether the star is west or east of the meridian. If it is west, $\tfrac{1}{2} t$ is in the first quadrant; if east, $\tfrac{1}{2} t$ is in the second. Applications of formula (38) may be found in §§ 81 and 82.

**8.** *Given the hour angle of a star, required its right ascension, and vice versa; the sidereal time in both cases being known.*

In Fig. 1, for any star $O$ we have

$VM$ = right ascension of star = $a$,
$MQ$ = hour angle of star = $t$,
$VQ$ = sidereal time = $\theta$.

Then
$$a = \theta - t, \qquad (40)$$
and
$$t = \theta - a, \qquad (41)$$

which effect the transformations.

TRANSFORMATION OF COÖRDINATES 13

Applications of (40) and (41) are numerous throughout the book.

**9.** *Given the right ascension and declination of a star, required its longitude and latitude, and vice versa.*

The transformation formulæ are obtained by applying the general equations (1), (2) and (3) to the triangle $POP'$, Fig. 1, in which

$$OP = 90° - \delta, \quad OP' = 90° - \beta, \quad OP'P' = 90° + \alpha,$$
$$OP'P = 90° - \lambda, \quad PP' = \text{obliquity of ecliptic} = \epsilon.$$

In order to adapt the resulting equations to logarithmic computation, assume

$$f \sin F = \sin \delta, \qquad (42)$$
$$f \cos F = \cos \delta \sin \alpha, \qquad (43)$$

and we shall obtain

$$\tan F = \frac{\tan \delta}{\sin \alpha}, \qquad (44)$$

$$\tan \lambda = \frac{\cos (F - \epsilon) \tan \alpha}{\cos F}, \qquad (45)$$

$$\tan \beta = \tan (F - \epsilon) \sin \lambda. \qquad (46)$$

The computations may be partially checked by the equation

$$\cos \delta \sin \alpha \sec F \cos (F - \epsilon) \operatorname{cosec} \lambda \sec \beta = 1, \qquad (47)$$

which is derived without difficulty from the transformation formulæ.

*Example.* The coördinates of *Regulus* on 1891 March 13 are

$$\alpha = 150° 38' 43''.5, \quad \delta = + 12° 29' 56''.4.$$

What are the corresponding longitude and latitude?

The necessary value of $\epsilon$, furnished by the American Ephemeris, page 278, is 23° 27' 16''.0. The resulting coördinates are

$$\lambda = 148° 19' 3''.1, \quad \beta = + 0° 27' 40''.5.$$

**10.** *Given the right ascensions and declinations of two stars, required the distance between them.*

Let the coördinates of the stars be $a'$, $\delta'$, and $a''$, $\delta''$, and $d$ the required distance. In the spherical triangle whose vertices are at the two stars and the pole, the sides are $90° - \delta'$, $90° - \delta''$ and $d$, and the angle at the pole is $a'' - a'$. Let $B'$ represent the angle opposite $90° - \delta'$. If in (1), (2) and (3) we put

$$a = d, \; B = B', \; b = 90° - \delta', \; c = 90° - \delta'', \; A = a'' - a',$$

they become

$$\cos d = \sin \delta' \sin \delta'' + \cos \delta' \cos \delta'' \cos (a'' - a'), \quad (48)$$
$$\sin d \cos B' = \sin \delta' \cos \delta'' - \cos \delta' \sin \delta'' \cos (a'' - a'), \quad (49)$$
$$\sin d \sin B' = \cos \delta' \sin (a'' - a'). \quad (50)$$

If $d$ can be determined from its cosine with sufficient precision, (48) will give the required distance; otherwise it should be determined from the tangent. If we assume

$$g \sin G = \cos \delta' \cos (a'' - a'), \quad (51)$$
$$g \cos G = \sin \delta', \quad (52)$$

we shall find that

$$\tan G = \cot \delta' \cos (a'' - a'), \quad (53)$$
$$\tan B' = \frac{\tan (a'' - a') \sin G}{\cos (\delta'' + G)}, \quad (54)$$
$$\tan d = \frac{\cot (\delta'' + G)}{\cos B'}. \quad (55)$$

(50) furnishes a partial check on the computations.

An application of these formulæ may be found in § 75, (*c*).

# CHAPTER II

## TIME

**11.** The passage of any point of the celestial sphere across the meridian of an observer is called the **transit**, or **culmination**, or **meridian passage** of that point. In one rotation of the sphere about its axis, every point of the sphere is twice on the meridian; once at *upper* culmination (above the pole), and once at *lower* culmination (below the pole). For an observer in the northern hemisphere, a star whose north polar distance is less than the latitude is constantly above the horizon, and both culminations are visible; a star whose south polar distance is less than the latitude is constantly below the horizon, and both culminations are invisible; and a star between these limits is visible at upper culmination, but invisible at the lower. For an observer in the southern hemisphere the first two cases are reversed.

Three systems of time are required in the operations of practical astronomy: **sidereal, apparent** (or **true**) **solar** and **mean solar**.

A **sidereal day** is the interval of time between two successive transits of the true vernal equinox over the same meridian. The **sidereal time** at any instant is the hour angle of the vernal equinox at that instant. It is $0^h\ 0^m\ 0^s$ when the vernal equinox is on the meridian — this instant is called **sidereal noon** — and is reckoned through 24 hours. The sidereal time is also equal to the right ascension of the observer's meridian, since the right ascension of the meridian is equal to the hour angle of the vernal equinox.

It follows, then, that any star will be at upper culmination at the instant when the sidereal time is equal to the star's right ascension; and at lower culmination when the sidereal time differs 12 hours from the star's right ascension. The rotation of the earth on its axis is perfectly uniform; but owing to precession and nutation the vernal equinox has a minute and irregular motion to the west (amounting on the average to $0''.126$ per day): so that a sidereal day does not correspond exactly to one rotation of the earth, nor is its length absolutely uniform, but it is *sensibly* so.

An **apparent** (or **true**) **solar day** is the interval of time between two successive upper transits of the sun over the same meridian. The hour angle of the sun at any instant is the **apparent time** at that instant. It is reckoned from $0^h\ 0^m\ 0^s$ at noon — called **apparent noon** — through 24 hours. But the apparent day varies greatly in length, for two reasons, viz.:

First, — The earth moves in an ellipse with a variable velocity. Hence the sun's (apparent) eastward motion (in longitude) is variable.

Second, — The sun's (apparent) motion is in the ecliptic.

Hence the sun's motion in right ascension and hour angle is variable, and a clock cannot be rated to keep apparent time.

A convenient solar time is obtained in this way: Assume an imaginary body to move in the ecliptic with a uniform angular velocity such that it and the sun pass through perigee at the same instant. Assume a second imaginary body to move in the equator with a uniform angular velocity such that the two will pass through the vernal equinox at the same instant. The second body is called the **mean sun**.

A **mean solar day** is the interval of time between two successive upper transits of the mean sun over the same meridian. The hour angle of the mean sun is the **mean** time. It is reckoned from $0^h 0^m 0^s$ at noon — called **mean noon** — through 24 hours.

The difference between the apparent and mean time is called the **equation of time**. Its value is given in the American Ephemeris for the instants of Greenwich apparent and mean noon and Washington apparent noon, whence its value may be obtained for any other instant by interpolation.

The astronomical solar day begins at noon, whereas the day popularly used — called the **civil day** — begins at (the preceding) midnight. Thus, Feb. 1, $10^h$ A.M., civil reckoning, is Jan. $31^d 22^h$ astronomical mean time.

**12.** The interval of time between two successive passages of the mean sun through the mean vernal equinox — called a **tropical year** — was for the year 1800, according to Bessel, 365.24222 * mean solar days.

The number of sidereal days in this interval is 366.24222, since in that interval of time the mean sun moves eastward through about 360°, and therefore the vernal equinox during the year makes one more transit over any given meridian than the sun. Thus we have

$$365.24222 \text{ mean days} = 366.24222 \text{ sidereal days.}$$

Whence

$24^h$ mean time $= 24^h\ 3^m\ 56^s.555$ sidereal time,

$24^h$ sidereal time $= 23\ 56\ \ 4.091$ mean time.

From these equations it is found that the gain of sidereal time on mean time in one mean hour is $9^s.8565$; and in one sidereal hour, $9^s.8296$. These are the amounts by which the right ascension of the mean sun increases in one mean and one sidereal hour, respectively.

---

* The length of the tropical year is diminishing at the rate of about $0^s.6$ per century. This is due to the fact that the mean vernal equinox is moving westward at an *accelerated* rate, as will be seen later, from the last of equations (120).

18  PRACTICAL ASTRONOMY

CONVERSION OF TIME

**13.** In nearly every problem of practical astronomy it is necessary to convert the time at one place into the corresponding time at another place, or to convert the time in one system into the corresponding time in another system. By means of the data furnished in the Ephemeris this is readily done.

**14.** *To convert the time at one place into the corresponding time at another.*

Since every epoch of time is defined by an hour angle, the difference of time at two places is the difference of the two corresponding hour angles; and that is equal to the difference of the longitudes of the two places. Therefore, if the difference of longitude be added to the time at the western place the sum is the corresponding time at the eastern. If it be subtracted from the time at the eastern place the result is the time at the western.

*Example 1.* The Ann Arbor mean time is 1891 March $10^d\ 21^h\ 10^m\ 54^s.70$. What is the corresponding Greenwich mean time?

Ann Arbor mean time,          1891 March $10^d\ 21^h\ 10^m\ 54^s.70$
Longitude Ann Arbor, Amer. Ephem., p. 482,  + 5  34  55 .14
Greenwich mean time           1891 March 11   2  45  49 .84

*Example 2.* The Washington sidereal time is $0^h\ 23^m\ 17^s.10$. What is the corresponding Ann Arbor sidereal time?

Washington sidereal time,            $0^h\ 23^m\ 17^s.10$
Difference of longitude, Amer. Ephem., p. 482,   0  26  43 .10
Ann Arbor sidereal time,             23  56  34 .00

*Example 3.* The Ann Arbor apparent time is 1891 March $20^d\ 21^h\ 58^m\ 19^s.17$. What is the Berlin apparent time at the same instant?

Ann Arbor apparent time,    1891 March $20^d\ 21^h\ 58^m\ 19^s.17$
Difference of longitude,                 6  28  30 .05
Berlin apparent time,       1891 March 21   4  26  49 .22

**15.** *To convert apparent time at any place into mean time, and vice versa.*

The equation of time at the given instant is required. When this is applied with the proper sign to the one, it gives the other. If apparent time is given, convert it into Greenwich apparent time, and take the equation of time from page I of the given month in the Ephemeris. If mean time is given, convert it into Greenwich mean time, and take the equation of time from page II of the month.

In taking these and other data from the Ephemeris, care must be exercised in making the interpolations. Thus, let it be required to determine the equation of time at Greenwich apparent time 1891 Feb. $24^d\,10^h$. Its value for apparent noon is $+13^m\,25^s.52$, and the difference for one hour *at noon* of that day is $0^s.381$. The difference for one hour *at noon* the next day is $0^s.406$. The hourly difference is therefore variable, but we may assume the second difference to be constant. The change in the equation during the 10 hours after noon is ten times the *average* hourly change for the 10 hours; that is, since the second difference is constant, ten times the hourly change *at the middle period,* or at 5 hours after noon. The average hourly change is $0^s.386$, and the desired equation of time is

$$+13^m\,25^s.52 - 10 \times 0^s.386 = +13^m\,21^s.66.$$

*Example.* The Berlin mean time is 1891 Feb. $28^d\,0^h\,11^m\,20^s.60$. What is the apparent time?

| | |
|---|---|
| Berlin mean time, | 1891 Feb. $28^d\ 0^h\ 11^m\ 20^s.60$ |
| Longitude Berlin, | $-\ \ \ \ 0\ \ 53\ \ 34.91$ |
| Greenwich mean time, | Feb. 27  23  18 |

This is $23^h.30$ after Gr. mean noon Feb. 27, or $0^h.70$ before noon Feb. 28. In this and similar cases the interpolation should be made for the interval before noon.

| | |
|---|---|
| Equation of time, Gr. mean noon, Feb. 28, | $-\ 12^m\ 44^s.52$ |
| Change before noon, $0.70 \times 0^s.473$, | $0.33$ |
| Equation of time, | $-\ 12\ \ 44.85$ |
| Berlin apparent time, | 1891 Feb. $27^d\ 23^h\ 58\ \ 35.75$ |

**16.** *To convert a mean time interval into the equivalent sidereal interval, and vice versa.*

In § 12 it is shown that sidereal time gains $9^s.8565$ on mean time in one mean hour. The corresponding gain for any number of hours, minutes, and seconds, is tabulated in Table III of the appendix to the American Ephemeris. If this gain be added to the mean time interval, the sum is the equivalent sidereal interval.

The gain of sidereal time on mean time in one sidereal hour is $9^s.8296$. The corresponding gain for any number of hours, minutes, and seconds, is tabulated in Table II of the appendix to the American Ephemeris. If this gain be subtracted from the sidereal interval, the difference is the equivalent mean time interval.

*Example 1.* A mean time interval is $17^h\ 33^m\ 21^s.76$. Find the corresponding sidereal interval.

| | |
|---|---:|
| Mean time interval, | $17^h\ 33^m\ 21^s.76$ |
| Gain of sidereal on mean, Table III, | 2 53 .04 |
| Sidereal interval, | 17 36 14 .80 |

*Example 2.* A sidereal time interval is $17^h\ 36^m\ 14^s.80$. Find the corresponding mean time interval.

| | |
|---|---:|
| Sidereal interval, | $17^h\ 36^m\ 14^s.80$ |
| Gain of sidereal on mean, Table II, | 2 53 .04 |
| Mean time interval, | 17 33 21 .76 |

**17.** *To convert mean time into sidereal time.*

Mean time at any instant is the interval after mean noon. If this interval be converted into the equivalent sidereal interval and added to the sidereal time at noon, the sum will be the sidereal time required. The sidereal time at noon is equal to the right ascension of the mean sun at this instant. The Ephemeris gives on page II for the month the sidereal time, or the right ascension of the mean sun, at Greenwich mean noon, whence its right ascension at noon for a place whose longitude is $L$ may be obtained by applying the term $L \times 9^s.8565$,* from Table III of the appendix to the American Ephemeris.

*Example.* The Ann Arbor mean time is 1891 Feb. $20^d$ $11^h\ 45^m\ 20^s.40$. What is the equivalent sidereal time?

Right ascension mean sun at Gr. mean noon, Feb. 20,　$22^h\ \ 0^m\ 31^s.75$
Change in $5^h\ 34^m\ 55^s.14$, Table III,　$0\ \ 55\ .02$
Right ascension, or sid. time, at Ann Arbor mean noon,　$22\ \ \ 1\ \ 26\ .77$
Mean time interval after noon,　$11\ \ 45\ \ 20\ .40$
Gain of sidereal on mean time, Table III,　$1\ \ 55\ .87$
Equivalent sidereal interval after noon,　$11\ \ 47\ \ 16\ .27$
Sidereal time,　$9\ \ 48\ \ 43\ .04$

For Ann Arbor, and similarly for other stations, the quantity $L \times 9^s.8565 = 55^s.02$ is a constant, and is held in mind by the computer.

Likewise, the experienced computer writes down the four necessary quantities and combines them all in one addition, thus:

$22^h\ \ 0^m\ 31^s.75$
$55\ .02$
$11\ 45\ 20\ .40$
$1\ 55\ .87$
$9\ 48\ 43\ .04$

**18.** *To convert sidereal time into mean time.*

If the sidereal time at the preceding mean noon (formed as before) be subtracted from the given time, the result is the sidereal interval after mean noon. This interval converted into the equivalent mean time interval is the mean time desired.

*Example.* On 1891 Feb. 20, the sidereal time at Ann Arbor is $9^h\ 48^m\ 43^s.04$. What is the mean time?

---

* It must be remembered that for a station east of Greenwich the quantity $L \times 9^s.8565$ is negative.

Right ascension mean sun at Gr. mean noon, Feb. 20,   $22^h$   $0^m$  $31^s.75$
Change in $5^h$ $34^m$ $55^s.14$, Table III,                          0   55 .02
Right ascension, or sid. time, at Ann Arbor mean noon, 22   1   26 .77*
The given sidereal time,                                    9  48  43 .04
Sidereal interval after mean moon, Feb. 20,                11  47  16 .27
Gain of sidereal on mean time, Table II,                           1  55 .87
Ann Arbor mean time,                         Feb. $20^d$ 11  45  20 .40

It should be noted that in the original statement of this example, the date 1891 Feb. 20 is the astronomical mean solar date. The observer should always record this date with care, especially in the case of observations taken near noon, as ambiguity may otherwise arise. It would be well in such cases, as indeed in all cases, to record also the civil day of the week.† Thus the statement "A daylight meteor was observed at Ann Arbor, 1891 Dec. 21, at sidereal time $18^h$ $2^m$ $30^s$," is ambiguous to the extent of one sidereal day, since on that solar day the sidereal time was *twice* equal to $18^h$ $2^m$ $30^s$. The record may refer to a phenomenon observed just after noon of Monday or just before noon of Tuesday. If the record were written "Monday, 1891 Dec. 20" there would be no uncertainty.

---

* This quantity is to be subtracted from the one directly following.

† Inasmuch as one may easily record an erroneous day of the month, many observers have the admirable practice of *beginning their records with the day of the week.* Thus, " Wednesday, 1801 July 8 " suffices for observations made on Wednesday afternoon, or for *continuous* observations throughout Wednesday night; but an *isolated* observation made the next morning may be headed, "Thursday morning, 1801 July 8."

# CHAPTER III

## CORRECTION OF OBSERVATIONS

**19.** The observed directions of all bodies in the solar system are sensibly different for observers at different places on the earth's surface. These differences must be allowed for before observations made at different places can be compared. This is accomplished by reducing all observations to the center of the earth, to which point the data of the Ephemeris refer. A knowledge of the form and size of the earth is therefore indispensable.

### FORM AND DIMENSIONS OF THE EARTH

**20.** Geodetic measurements, combined with astronomical observations, have shown that the earth is very nearly an oblate spheroid whose minor axis coincides with the polar axis.

Let $QPQ'P'$ be an elliptical section of the spheroid made by the meridian of an observer at $O$; $A$ the center of the earth; $NS$ the horizon; and let

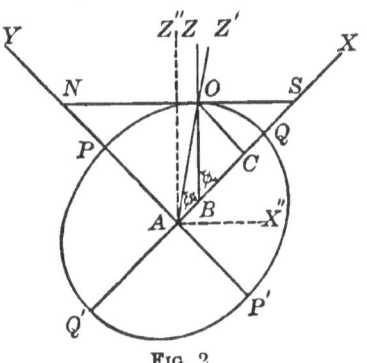

Fig. 2

$a$ = semi-major axis of ellipse
$\quad = AQ$,
$b$ = semi-minor axis of ellipse
$\quad = AP$,
$\phi$ = geographical latitude* of $O$
$\quad = OBQ$,
$\phi'$ = geocentric latitude of $O$
$\quad = OAQ$,
$\rho$ = radius of earth at $O = AO$,
$\phi' - \phi$ = reduction to geocentric latitude $= AOB$,
$x, y$ = rectangular coördinates of
$\quad O = AC, CO$.

---

\* It frequently happens, especially in mountainous regions, that the plumb-line is not normal to the theoretical ellipsoidal surface of the earth,

From a discussion of all available observations, Bessel found
$$a = 3962.802 \text{ miles}, \qquad b = 3949.555 \text{ miles};$$
and therefore, for the eccentricity of a meridional section, $QPQ'P'$,
$$e = 0.0816967.$$

**21.** *Given the geographical latitude of a point on the earth's surface, required the corresponding geocentric latitude.*

The equation of the ellipse (Fig. 2) is
$$\frac{x^2}{a^2} + \frac{y^2}{b^2} = 1. \tag{56}$$

Differentiating, and substituting
$$\tan \phi = -\frac{dx}{dy}, \qquad \tan \phi' = \frac{y}{x},$$
we obtain the desired relation
$$\tan \phi' = \frac{b^2}{a^2} \tan \phi = (1 - e^2) \tan \phi. \tag{57}$$

The reduction to the geocentric latitude, $\phi' - \phi$, can be expressed in terms of $\phi$. If the equation
$$\tan x = p \tan y,$$
which is identical in form with (57), be developed in series it becomes [*Chauvenet's Plane Trig.*, § 254]
$$x - y = q \sin 2y + \tfrac{1}{2} q^2 \sin 4y + \tfrac{1}{3} q^3 \sin 6y + \cdots,$$
in which
$$q = \frac{p-1}{p+1}.$$

---

owing to the fact that the local irregularities of surface and of density become appreciable. In such cases, the zenith determined by the plumb-line will not coincide with the theoretical zenith. Consequently the latitude and longitude, determined astronomically, will differ from the latitude and longitude determined geodetically. The geodesist has to deal with both systems, but the astronomer uses only the former.

Substituting from (57) the values corresponding to $x$, $y$, and $p$, and dividing by sin $1''$ in order to express the result in seconds of arc, we obtain the practically rigorous formula

$$\phi' - \phi = -690''.65 \sin 2\phi + 1''.16 \sin 4\phi. \tag{58}$$

**22.** *To find the radius of the earth for a given latitude.*

Substituting $x = \rho \cos \phi'$ and $y = \rho \sin \phi'$ in (56) and eliminating $b$ by (57), we obtain

$$\rho = a\sqrt{\frac{\cos \phi}{\cos(\phi - \phi') \cos \phi'}}. \tag{59}$$

In using this equation make $a = 1$, since the equatorial radius is taken as the unit.

The values of $\phi' - \phi$ and of $\rho$ for the positions of the principal observatories are given on pp. 482–485 of the American Ephemeris for 1891.

Formulæ (58) and (59) give the correct values of $\phi' - \phi$ and $\rho$ *at sea level*. It is evident that the altitude of the observer above sea level must be taken into account. The slight corrections thus rendered necessary may be computed from elementary principles of trigonometry.

### PARALLAX

**23.** The **geocentric** or **true place** of a star is that in which it would be seen by an observer at the center of the earth. The **apparent*** or **observed place** is that in which it is seen by the observer on the surface of the earth. The **parallax** of a star is the difference between its true and apparent places. It may also be defined as the angle at the star subtended by the radius of the earth drawn to the point of observation. This angle is approximately a

---

* The terms *true* and *apparent* are used in a relative sense only. In reference to parallax, the true place is the place corrected for parallax. In reference to refraction, the apparent place is affected by refraction, the true place is corrected for refraction; and similarly in other subjects.

maximum for an observer at a given place when the star is seen in his horizon. It is then called the **horizontal parallax**. When the observer is at a place on the earth's equator this angle is called the **equatorial horizontal parallax**.

**24.** *To find the equatorial horizontal parallax of a body.*

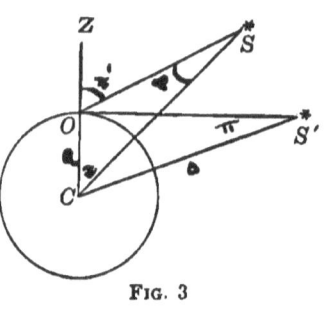

Fig. 3

In Fig. 3 let $S'$ be a body in the horizon of a point $O$ on the earth's equator. Then if

$a$ = equatorial radius of the earth = $CO$,

$\Delta$ = body's distance from the earth's center = $CS'$,

$\pi$ = equatorial horizontal parallax = $CS'O$,

we have $\qquad \sin \pi = \dfrac{a}{\Delta}.$ (60)

The astronomical unit of distance is the mean distance of the earth from the sun. Using (60) with $a$ and $\Delta$ expressed in terms of this unit, the American Ephemeris tabulates the values of $\pi$ for the moon [page IV of the month] and the planets [pp. 218–249]; employing values of $a$ and the earth's mean distance from the sun such that the sun's mean equatorial horizontal parallax is $8''.848$.

Recent researches have shown that $8''.80$ is probably a much more correct value of the sun's mean equatorial horizontal parallax, and the superintendents of the principal astronomical annuals have agreed to use that value in computing ephemerides from about the year 1900.

**25.** *To find the parallax in zenith distance, the earth being regarded as a sphere.*

In Fig. 3 let

$z'$ = apparent zenith distance of a star $S = ZOS$,

$z$ = true zenith distance = $ZCS$,

$p$ = parallax in zenith distance = $CSO = z' - z$.

PARALLAX 27

Then from the triangle $COS$

$$\frac{\sin p}{\sin z'} = \frac{a}{\Delta} = \sin \pi,$$

or

$$\sin p = \sin \pi \sin z'. \quad (61)$$

For all bodies except the moon we can write

$$p = \pi \sin z'. \quad (62)$$

For the sun we have, from (60),

$$\pi = \frac{8''.8}{\Delta};$$

and (62) becomes

$$p = \frac{8''.8}{\Delta} \sin z'. \quad (63)$$

The values of $\log \Delta$ are tabulated in the Ephemeris, page III of the month.

In the case of observations made with a sextant, surveyor's transit, or other similar instrument, we may assume $\Delta$ equal to unity, and take the required value of $p$ from the following table computed from

$$p = 8''.8 \sin z': \quad (64)$$

| $z'$ | $p$ | $z'$ | $p$ |
| --- | --- | --- | --- |
| 0° | 0″.0 | 50° | 6″.7 |
| 10 | 1 .5 | 60 | 7 .6 |
| 20 | 3 .0 | 70 | 8 .3 |
| 30 | 4 .4 | 80 | 8 .7 |
| 40 | 5 .7 | 90 | 8 .8 |

For refined observations (61) is not sufficiently exact, and recourse must be had to formulæ which consider the earth as a spheroid.

**26.** *Given the true zenith distance and azimuth of a star, required its apparent zenith distance and azimuth, the earth being regarded as a spheroid.*

Let the star be referred to a system of rectangular axes whose origin is at the point of observation, the positive axis of $x$ being directed to the south point, the positive axis of $y$ to the west point and the positive axis of $z$ to the zenith. Let

$X'$, $Y'$, $Z'$ = the rectangular coördinates of the star,
$\Delta'$ = the star's distance from the observer,
$A'$ = its apparent azimuth,
$z'$ = its apparent zenith distance.

Then
$$X' = \Delta' \sin z' \cos A',$$
$$Y' = \Delta' \sin z' \sin A',$$
$$Z' = \Delta' \cos z'.$$

Again, let the star be referred to a second system of rectangular axes parallel to the first, the origin being at the center of the earth. Let

$X$, $Y$, $Z$ = the rectangular coördinates of the star,
$\Delta$ = the star's distance from the origin,
$A$ = its true azimuth,
$z$ = its true zenith distance.

Then
$$X = \Delta \sin z \cos A,$$
$$Y = \Delta \sin z \sin A,$$
$$Z = \Delta \cos z.$$

Let the coördinates of the point of observation referred to the second system be $X''$, $Y''$, $Z''$. From Fig. 2 it is seen that

$$X'' = \rho \sin (\phi - \phi'), \quad Y'' = 0, \quad Z'' = \rho \cos (\phi - \phi').$$

Now
$$X' = X - X'', \quad Y' = Y - Y'', \quad Z' = Z - Z'',$$

and therefore

$$\left. \begin{array}{l} \Delta' \sin z' \cos A' = \Delta \sin z \cos A - \rho \sin (\phi - \phi'), \\ \Delta' \sin z' \sin A' = \Delta \sin z \sin A, \\ \Delta' \cos z' \phantom{XXX} = \Delta \cos z \phantom{XXX} - \rho \cos (\phi - \phi'). \end{array} \right\} \quad (65)$$

These equations completely determine $\Delta'$, $z'$ and $A'$, and therefore the parallax $z' - z$ and $A' - A$. It is better, however, to transform them so that the parallax can be computed directly. For this purpose, divide the equations through by $\Delta$ and put

$$f = \frac{\Delta'}{\Delta};$$

also substitute from (60), $a$ being unity,

$$\sin \pi = \frac{1}{\Delta},$$

and we have

$$f \sin z' \cos A' = \sin z \cos A - \rho \sin \pi \sin (\phi - \phi'), \quad (66)$$
$$f \sin z' \sin A' = \sin z \sin A, \quad (67)$$
$$f \cos z' = \cos z - \rho \sin \pi \cos (\phi - \phi'). \quad (68)$$

From (66) and (67) we obtain

$$f \sin z' \sin (A' - A) = \rho \sin \pi \sin (\phi - \phi') \sin A, \quad (69)$$
$$f \sin z' \cos (A' - A) = \sin z - \rho \sin \pi \sin (\phi - \phi') \cos A. \quad (70)$$

Putting

$$m = \frac{\rho \sin \pi \sin (\phi - \phi')}{\sin z}, \quad (71)$$

(69) and (70) give

$$\tan (A' - A) = \frac{m \sin A}{1 - m \cos A}. \quad (72)$$

Multiplying (69) by $\sin \tfrac{1}{2} (A' - A)$ and (70) by $\cos \tfrac{1}{2} (A' - A)$, adding the products and dividing by $\cos \tfrac{1}{2} (A' - A)$, we obtain

$$f \sin z' = \sin z - \rho \sin \pi \sin(\phi - \phi') \frac{\cos \tfrac{1}{2} (A' + A)}{\cos \tfrac{1}{2} (A' - A)}. \quad (73)$$

Let us assume

$$\tan \gamma = \tan (\phi - \phi') \frac{\cos \tfrac{1}{2} (A' + A)}{\cos \tfrac{1}{2} (A' - A)}; \quad (74)$$

then

$$f \sin z' = \sin z - \rho \sin \pi \cos (\phi - \phi') \tan \gamma. \quad (75)$$

This combined with (68) gives

$$f \sin (z' - z) = \rho \sin \pi \cos (\phi - \phi') \frac{\sin (z - \gamma)}{\cos \gamma}, \qquad (76)$$

$$f \cos (z' - z) = 1 - \rho \sin \pi \cos (\phi - \phi') \frac{\cos (z - \gamma)}{\cos \gamma}. \qquad (77)$$

Assume

$$n = \frac{\rho \sin \pi \cos (\phi - \phi')}{\cos \gamma}, \qquad (78)$$

and we have

$$\tan (z' - z) = \frac{n \sin (z - \gamma)}{1 - n \cos (z - \gamma)}. \qquad (79)$$

Formulæ (71) and (72) rigorously determine the parallax in azimuth, and (74), (78) and (79) the parallax in zenith distance. We may abbreviate the computation by writing (74) in the form

$$\gamma = (\phi - \phi') \cos A, \qquad (80)$$

which is in all cases sufficiently exact.

**27.** *Given the apparent zenith distance and azimuth of a body, required its true zenith distance and azimuth, the earth being regarded as a spheroid.*

From (68) and (75) we obtain

$$\sin (z' - z) = \frac{\rho \sin \pi \cos (\phi - \phi') \sin (z' - \gamma)}{\cos \gamma},$$

for which, since $\phi - \phi'$ and $\gamma$ are small angles, we can write

$$\sin (z' - z) = \rho \sin \pi \sin (z' - \gamma), \qquad (81)$$

in which $\gamma$ is given without sensible error by

$$\gamma = (\phi - \phi') \cos A'. \qquad (82)$$

We obtain from (66) and (67)

$$\sin (A' - A) = \frac{\rho \sin \pi \sin (\phi - \phi') \sin A'}{\sin z}, \qquad (83)$$

in which the value of $z$ is found by the solution of (81). Formulæ (82), (81) and (83) completely solve the problem. For all known bodies save the moon we may write

PARALLAX

$$z' - z = \rho \pi \sin(z' - \gamma), \quad (84)$$
$$A' - A = \rho \pi \sin(\phi - \phi') \sin A' \csc z'. \quad (85)$$

An application of the formulæ of this section will be found in § 89, in the determination of longitude by lunar distances.

**28.** *To find the parallax of a body in right ascension and declination.* Let

$a =$ the body's geocentric right ascension,
$\delta =$ " " " declination,
$t =$ " " " hour angle,
$\Delta =$ " " " distance,
$a' =$ " " apparent right ascension,
$\delta' =$ " " " declination,
$t' =$ " " " hour angle,
$\Delta' =$ " " " distance,
$\theta =$ the observer's sidereal time.

By methods similar to those used in developing equations (72), (74) and (79), we may obtain the corresponding equations,

$$\tan(a - a') = \frac{\rho \sin \pi \cos \phi' \sin t}{\cos \delta - \rho \sin \pi \cos \phi' \cos t}, \quad (86)$$

$$\tan \gamma = \frac{\tan \phi' \cos \tfrac{1}{2}(a - a')}{\cos[\theta - \tfrac{1}{2}(a + a')]}, \quad (87)$$

$$\tan(\delta - \delta') = \frac{\rho \sin \pi \sin \phi' \sin(\gamma - \delta)}{\sin \gamma - \rho \sin \pi \sin \phi' \cos(\gamma - \delta)} \quad (88)$$

These rigorously determine the parallax in right ascension, $a - a'$, and the parallax in declination, $\delta - \delta'$, when the geocentric coördinates are the known quantities. If the apparent coördinates $a'$, $\delta'$ and $t'$ have been obtained by observation, and $a$, $\delta$ and $t$ are unknown, we substitute $a'$, $\delta'$ and $t'$ for $a$, $\delta$ and $t$ in the second members, and solve. The resulting approximate values of the parallax furnish nearly correct values of $a$, $\delta$ and $t$. Employing these in a second solution of the equations we shall obtain sufficiently exact values of the parallax.

**29.** For all known bodies except the moon the values of $\pi$, $a - a'$ and $\delta - \delta'$ will be very small, and we may write (86), (87) and (88), without sensible error, in the form

$$a - a' = \frac{8''.8\, \rho \cos \phi' \sin t}{\Delta \cdot \cos \delta}, \tag{89}$$

$$\tan \gamma = \frac{\tan \phi'}{\cos t}, \tag{90}$$

$$\delta - \delta' = \frac{8''.8\, \rho \sin \phi' \sin (\gamma - \delta)}{\Delta \cdot \sin \gamma}, \tag{91}$$

in which $\Delta$ is expressed in terms of the astronomical unit of distance. These formulæ will determine the parallax satisfactorily also if $t$ and $\delta$ are replaced by $t'$ and $\delta'$, for which case an application of them will be found in § 155.

At the fixed observatories it is customary to construct tables which greatly facilitate the computation of parallaxes. The equations (89) and (91) may be written

$$(a - a')\, \Delta = 8''.8\, \rho \cos \phi' \sin t' \sec \delta', \tag{92}$$

$$(\delta - \delta')\, \Delta = 8''.8\, \rho \sin \phi' \csc \gamma \sin (\gamma - \delta'). \tag{93}$$

The second members of these equations are the parallaxes in right ascension and declination of an imaginary body at distance unity when observed, at a given station $(\rho, \phi')$, in the direction $t'$, $\delta'$. They are called **parallax factors**. Their values are generally computed and tabulated, at a given observatory, for every $10^m$ of hour angle and every degree of declination. When a body is observed at any hour angle and declination, the corresponding parallax factors may be obtained by interpolation from the tables. The parallaxes, $a - a'$ and $\delta - \delta'$, may then be determined by dividing the parallax factors by the distance $\Delta$ of the body, as will be seen from (92) and (93). An application of these formulæ will be found in § 154.

### REFRACTION

**30.** It is shown in Optics that when a ray of light passes obliquely from one transparent medium into another of greater density, it is refracted from its original direction according to the following laws :

(a) The incident ray, the normal to the surface which separates the two media at the point of incidence, and the refracted ray, lie in the same plane.

(b) The sines of the angles of incidence and refraction are inversely as the indices of refraction of the two media.

A ray of light coming from a star to an observer is assumed to travel in a straight line until it reaches the upper limit of the earth's atmosphere. It then passes continually from a rarer to a denser medium until it reaches the earth's surface. If we regard the earth as a sphere, it follows from (a) and (b) that the path of the ray is a curve whose direction constantly approaches the center of the earth.

Let Fig. 4 represent a section of the earth and its atmosphere made by a vertical plane passing through the point of observation $O$ and a star $S$. The path of the ray, $S\,a\,b\,..\,n\,..\,O$, lies wholly in this plane and is concave towards the earth. The *apparent* direction of the star is $OS'$, a tangent to the curve at the point of observation. The *true* direction is that of a straight line joining $O$ and $S$. The difference of these directions is the **refraction**. It appears that refraction increases the altitude, and decreases the zenith distance, of a star, but in general does not affect its azimuth.*

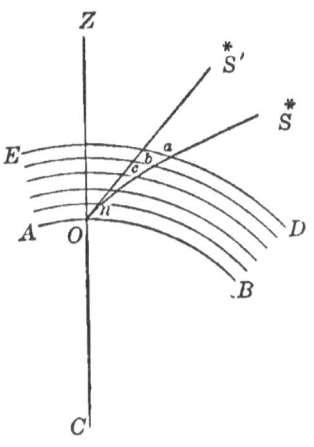

FIG. 4

---

* It is known that appreciable deviations in the azimuth are sometimes produced by refraction, especially in observations made very near the horizon ; but as they are due to abnormal and unknown arrangements of the strata of air, there is unfortunately no direct method of eliminating them.

D

The amount of the refraction depends upon the density of the air, which is a function of the atmospheric pressure and temperature. Our knowledge of the state of the atmosphere is very imperfect. The theory of refraction is complex and tedious, refraction tables to be reliable must be largely empirical, and we shall not attempt an investigation of the subject.

The *Pulkowa Refraction Tables* given in the Appendix, TABLE I, are based on the formula

$$r = \mu \tan z \, (BT)^A \gamma^\lambda \sigma^i, \qquad (94)$$

in which $z$ is the *apparent* zenith distance, $\mu$, $A$, $\lambda$, and $\sigma$ are functions of the apparent zenith distance, $B$ depends on the reading of the **barometer**, $T$ depends on the temperature of the column of mercury as indicated by the **attached*** **thermometer**, $\gamma$ depends on the temperature of the atmosphere as indicated by the **external thermometer**, $i$ depends on the time of the year, and $r$ is the refraction in seconds of arc.

For logarithmic computation (94) takes the form

$$\log r = \log \mu + \log \tan z + A \,(\log B + \log T) + \lambda \log \gamma + i \log \sigma. \quad (95)$$

Observations should not be made at a greater zenith distance than 82° 30′, beyond which the amount of the refraction is uncertain. We can compute an approximate value of the refraction, however, by means of the *Supplement* to TABLE I, which tabulates the values of $\log \mu \tan z$.

*Example.* Given the apparent zenith distance 81° 11′ 0″, Barom. 29.420 inches, Attached Therm. + 46°.5 F., External Therm. + 22°.3 F., time May 20, required the true zenith distance.

| | | | |
|---|---|---|---|
| $\log B$ | − 0.00260 | $\log \mu$ | 1.74132 |
| $\log T$ | − 0.00055 | $\log \tan z$ | 0.80937 |
| $\log BT$ | − 0.00315 | $A \log BT$ | 9.99683 |
| $A$ | 1.0053 | $\lambda \log \gamma$ | 0.02456 |
| $\log \gamma$ | + 0.02337 | $i \log \sigma$ | 9.99995 |
| $\lambda$ | 1.0510 | $\log r$ | 2.57203 |
| $\log \sigma$ | 0.00026 | $r$ | 6′ 13″.3 |
| $i$ | − 0.21 | | |

---

* That is, attached to the barometer.

The true zenith distance is therefore 81° 17′ 13″.3.

If the true zenith distance is given and the apparent zenith distance is required, an approximate value of the latter is first found by applying the **mean refraction**, TABLE II, Appendix, to the true zenith distance, and then the refraction is given by (94) as before.

TABLE II is constructed from (94) for a mean state of the atmosphere, viz.: Barom. 29.5 inches, Att. Therm. 50°.0, and Ext. Therm. 50°.0. The factor $\sigma^i$ is neglected.

In case no tables are available an approximate value of the refraction is given by

$$r = \frac{983\,b}{460 + t} \tan z, \qquad (96)$$

in which $b$ is the barometer reading in inches, $t$ the temperature of the atmosphere in degrees Fahr., and $z$ the apparent zenith distance.* For zenith distances less than 75° it represents the Pulkowa refractions within a second of arc, except for extreme states of the atmosphere. It is especially convenient for field work in which an aneroid barometer is used.

When the barometer and thermometer have not been read a roughly approximate value of the refraction is given by

$$r = K \tan z, \qquad (97)$$

in which $K$ is the value of the refraction at zenith distance 45° for the *mean* barometer and thermometer readings at the place of observation; but for large zenith distances and extreme states of the atmosphere it cannot be used safely. The value of $K$ for sea-level stations in the temperate zones is about 58″.

---

* This formula is due to Professor Comstock: *The Sidereal Messenger*, April, 1890.

## REFRACTION IN RIGHT ASCENSION AND DECLINATION

**31.** The change in zenith distance due to refraction gives rise to corresponding changes in right ascension and declination. We know the general relations existing between these coördinates, whence the relations existing between their increments may be found by differentiation. From (7), $\delta$ and $z$ being the only variables, we have

$$\cos \delta\, d\delta = -(\sin z \sin \phi + \cos z \cos \phi \cos A)dz,$$

which reduces by means of (24) to

$$d\delta = -\cos q\, dz. \tag{98}$$

Differentiating (30), regarding $z$, $\delta$ and $t$ as variables, we obtain

$$-\sin z\, dz = (\cos \delta \sin \phi - \sin \delta \cos \phi \cos t)\, d\delta - \cos \delta \cos \phi \sin t\, dt,$$

which by (31), (32) and (98) reduces to

$$\cos \delta\, dt = \sin q\, dz. \tag{99}$$

But from (41) $dt = -d\alpha$. Making this substitution and replacing $dz$ by the refraction $r$, (98) and (99) become

$$d\delta = -r \cos q, \tag{100}$$
$$d\alpha = -r \sin q \sec \delta. \tag{101}$$

These corrections reduce from the apparent to the true values of $\alpha$ and $\delta$. If the true place is given and the apparent place is required, the signs of the corrections must be reversed.

To compute $r$ we must know $z$. If $z$ and $A$ are given, $q$ is determined by (29); if $t$ and $\delta$ are known, $q$ and $z$ are determined by (36) and (37).

### DIP OF THE HORIZON

**32.** At sea the altitudes of celestial objects are measured from the **visible sea horizon**. This is below the true

# DIP OF THE HORIZON

horizon by an amount depending on the elevation of the observer's eye above the surface of the sea.

Let Fig. 5 represent a section of the earth made by a vertical plane passing through the eye of an observer at $O$. $OH'$ is a line in the visible horizon, $OH$ is the corresponding line in the true horizon, and $HOH'$ is the **dip of the horizon**. Let

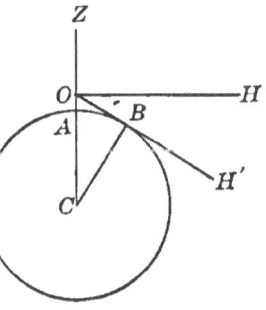

Fig. 5

$x$ = the height of the eye above the water in feet = $OA$,

$a$ = the radius of the earth in feet = $AC$,

$D$ = the dip of the horizon = $HOH'$ = $OCB$.

We may write

$$\tan D = \frac{OB}{CB} = \frac{\sqrt{2ax + x^2}}{a} = \sqrt{\frac{2x}{a} + \frac{x^2}{a^2}}. \qquad (102)$$

But $\dfrac{x^2}{a^2}$ is a very small quantity and may be neglected. Tan $D$ may be replaced by $D \tan 1''$. The apparent dip is affected by refraction. The amount of this refraction is uncertain, but an approximate value of the true dip is obtained by multiplying the apparent dip by the factor 0.92. The mean value of $a$ is 20888625 feet. Introducing these quantities in (102) it reduces to

$$D = 59'' \sqrt{x \text{ in feet}}, \qquad (103)$$

by which amount the measured altitude must be decreased.

A convenient rule, much used by navigators, follows approximately from (103), thus:

The dip in minutes of arc is expressed by the square root of the number of feet that the observer's eye is above the water. To illustrate, if the observer's eye is 30 feet above the water, the dip is very nearly $5'.5$.

The dip must in all cases be subtracted from the observed altitude in order to obtain the true altitude.

## SEMIDIAMETER

**33.** When we observe a celestial body having a well-defined disk, as in the case of the sun and moon, the measurements are made with reference to some point on the limb, and the position of the center is obtained by correcting the observation for the angular semidiameter of the body.

The geocentric semidiameters of the sun, moon and major planets are tabulated in the Ephemeris. The apparent semidiameter of the moon, however, is appreciably different for different altitudes, on account of its nearness to the earth, and its value must be determined.

**34.** *To find the apparent semidiameter of the moon.*

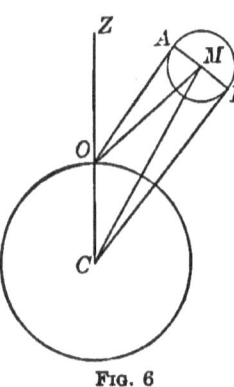

Fig. 6

Let Fig. 6 represent a section made by a plane passing through the observer $O$, the center of the moon $M$, and the center of the earth $C$, the earth being considered a sphere.* Let

$S$ = the moon's geocentric semidiameter
   = $MCB$,
$S'$ = the moon's apparent semidiameter
   = $AOM$,
$\Delta$ = the distance of the moon's center from the earth's center = $CM$,
$\Delta'$ = the distance of the moon's center from the observer = $OM$,
$\pi$ = the equatorial horizontal parallax of the moon,
$p$ = the parallax in zenith distance = $OMC$,
$z$ = the moon's true zenith distance = $ZCM$,
$z'$ = the moon's apparent zenith distance = $ZOM$.

Then we can write

$$\frac{\sin S'}{\sin S} = \frac{\Delta}{\Delta'} = \frac{\sin (z+p)}{\sin z} = \cos p + \frac{\cos z \sin p}{\sin z}.$$

---

* The maximum error produced by neglecting the eccentricity of the meridian even in the case of the moon never exceeds $0''.06$.

From (61)
$$\sin p = \sin \pi \sin z'; \qquad (104)$$
therefore
$$\sin S' = \sin S \left(\cos p + \sin \pi \cos z \frac{\sin z'}{\sin z}\right). \qquad (105)$$

(104) and (105) furnish very nearly an exact solution of the problem. For our purpose, and for all ordinary observations, we can write

$$S' = S(1 + \sin \pi \cos z). \qquad (106)$$

**35.** *To find the contraction of any semidiameter of the sun or moon, produced by refraction.*

The apparent disk of the sun or moon is not circular, since the refraction for the lower limb is greater than for the center, and that for the center is greater than for the upper limb. It will be sufficiently exact to assume the disk to be an ellipse whose center coincides with the center of the sun or moon.

The contraction of the vertical semidiameter is found by taking the difference of the refractions for the center and the upper or lower limb.

The contraction of the horizontal semidiameter for all zenith distances less than 85° is *very nearly constant* and equal to about $0''.25$. For our purpose it may be neglected, and we shall not investigate the subject.

The contraction of any semidiameter making an angle $q$ with the vertical semidiameter is readily obtained from the properties of the ellipse. Thus let

$a$ = the horizontal semidiameter,
$b$ = the vertical semidiameter,
$S''$ = the inclined semidiameter,

and we have
$$\frac{x^2}{a^2} + \frac{y^2}{b^2} = 1,$$
$$S'' \sin q = x,$$
$$S'' \cos q = y;$$

whence
$$S'' = \frac{ab}{\sqrt{a^2 \cos^2 q + b^2 \sin^2 q}}. \qquad (107)$$

## ABERRATION

**36.** The observed direction of a star differs from its true direction in consequence of the motion of the observer in space. The ratio of the velocity of light to the velocity of the observer is finite, and a telescope changes its position appreciably while a ray of light is passing from the objective to the eyepiece.

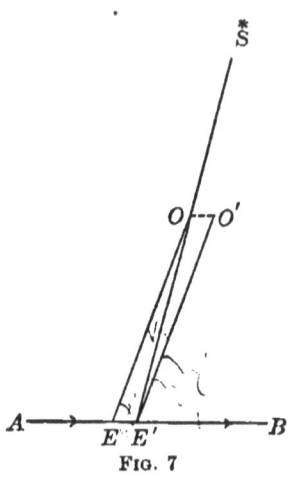

FIG. 7

In Fig. 7 let $O$ be the center of the objective and $E$ the center of the eyepiece of a telescope at the instant when a ray from a star $S$ reaches the point $O$. If $OE'$ represent the direction and velocity of the ray, and $AB$ represent the direction of the observer's motion and $EE'$ his velocity, the telescope will be in the position $O'E'$ when the ray reaches $E'$. While the true direction of the star is $E'O$, the apparent direction is $E'O'$. The change in the apparent direction, $OE'O'$, is called the aberration. The star is apparently displaced toward that point of the celestial sphere which the observer is momentarily approaching. To find the amount of this displacement let (Fig. 7)

$\gamma = BE'O$ = the angle between the true direction of the star and the line of the observer's motion,

$\gamma' = BE'O'$ = the angle between the apparent direction of the star and the line of the observer's motion,

$d\gamma = \gamma - \gamma'$ = the correction for aberration in the plane of the star and the observer's motion,

$V = OE'$ = the velocity of light,

$v = EE'$ = the velocity of the observer.

# ABERRATION

Then from the triangle $EOE''$ we have

$$\sin(\gamma - \gamma') = \sin d\gamma = \frac{v}{V}\sin\gamma'.$$

But $d\gamma$ is always very small and we can write, without sensible error,

$$d\gamma = \frac{v}{V \sin 1''}\sin\gamma, \qquad (108)$$

which determines the correction for aberration when $v$, $V$ and $\gamma$ are known.

**37.** The velocity of the observer is made up of three parts: those due to the motion of the solar system as a whole, to the annual motion of the earth in its orbit, and to the diurnal rotation of the earth. The first need not be considered, since it affects the apparent place of a star by a constant quantity. The second gives rise to **annual aberration**, which will be referred to in CHAPTER IV. The third gives rise to the **diurnal aberration**. This is a function of the observer's position on the earth, and will be treated as a correction to be applied to observed coördinates.

**38.** *To find the diurnal aberration in hour angle and declination.*

In Fig. 8 let $SENW$ be the horizon, $EQW$ the equator, $L$ the earth, $O$ a star whose hour angle is $t$ and declination $\delta$, $EOW$ a great circle through $O$ and the east point of the horizon. Owing to the diurnal rotation of the earth the observer is moving directly toward the east point, and therefore the star's apparent place is shifted eastward in the plane $EOW$ to some point $O'$. The aberration in this plane is $OO'$, whose value is given by (108). It

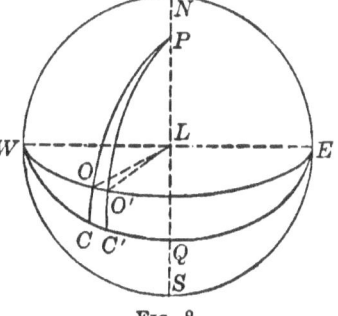

Fig. 8

only remains to find the corresponding change in hour angle $CC'$, and in declination $CO - C''O'$.

In the triangle $ECO$ we have

$$CO = \delta, \quad CE = 90° + t, \quad ECO = 90°.$$

Now let

$$OE = \gamma, \quad CEO = \omega, \quad C'C = dt, \quad CO - C'O' = d\delta,$$

and we can write

$$\sin \delta = \quad \sin \gamma \sin \omega, \qquad (109)$$
$$\sin t \cos \delta = - \cos \gamma, \qquad (110)$$
$$\cos t \cos \delta = \quad \sin \gamma \cos \omega. \qquad (111)$$

(110) and (111) give by differentiation, $t$, $\delta$ and $\gamma$ being variables,

$$- \sin t \sin \delta \, d\delta + \cos t \cos \delta \, dt = \sin \gamma \, d\gamma,$$
$$- \cos t \sin \delta \, d\delta - \sin t \cos \delta \, dt = \cos \gamma \cos \omega \, d\gamma.$$

Eliminating $d\delta$ and then $dt$, we obtain

$$\cos \delta \, dt = \quad (\sin \gamma \cos t - \cos \gamma \sin t \cos \omega) \, d\gamma,$$
$$\sin \delta \, d\delta = - (\sin \gamma \sin t + \cos \gamma \cos t \cos \omega) \, d\gamma,$$

which, by means of (109), (110) and (111), reduce to

$$dt = \quad \cos t \sec \delta \frac{d\gamma}{\sin \gamma}, \qquad (112)$$

$$d\delta = - \sin t \sin \delta \frac{d\gamma}{\sin \gamma}. \qquad (113)$$

The value of the factor $\dfrac{d\gamma}{\sin \gamma}$ is given by (108) from the known values of $v$ and $V$. For an observer at the earth's equator it is $0''.31$; in latitude $\phi$ it is $0''.31 \cos \phi$. Substituting this value in (112) and (113) we obtain

$$dt = + 0''.31 \cos \phi \cos t \sec \delta, \qquad (114)$$
$$d\delta = - 0.31 \cos \phi \sin t \sin \delta, \qquad (115)$$

which are the corrections to be applied to the observed hour angle and declination.

When the star is observed on the meridian, $t = 0$, and (114) and (115) become

$$dt = 0''.31 \cos \phi \sec \delta, \qquad (116)$$
$$d\delta = 0 . \qquad (117)$$

**39.** *To find the diurnal aberration in azimuth and altitude.*

The problem is identical with that in § 38 save that the horizon is the plane of reference, instead of the equator. If in (114) and (115) we replace $t$ by $A$ and $\delta$ by $h$ we obtain the desired corrections,

$$dA = + 0''.31 \cos \phi \cos A \sec h, \qquad (118)$$
$$dh = - 0.31 \cos \phi \sin A \sin h, \qquad (119)$$

which are the corrections to be applied to the observed azimuth and altitude.

### SEQUENCE AND DEGREE OF CORRECTIONS

**40.** In applying the corrections considered in this chapter it is necessary that a proper sequence be followed.

*In all cases* the refraction must be applied first, its amount being obtained by the methods of § 30 and § 31.

Except in a few cases the diurnal aberration may be neglected.

Observations on the sun or moon refer to points on the limb. They must be reduced to the center. In the case of the moon the reduction is made by formulæ (106) and (107); of the sun, by (107).

The parallax is now determined by the methods of §§ 25–29. It is wholly inappreciable for the stars.

The degree of refinement to which these corrections should be carried, can be stated only in a general way. Usually it is sufficient to compute the corrections to one order of units lower than that to which the observations have been made. Thus, in reducing an observation made with a sextant reading to $10''$, the corrections should be computed to the nearest second. If the mean of a large number of sextant readings is employed, it is advisable to carry the corrections to tenths of a second; and similarly in other cases.

# CHAPTER IV

PRECESSION — NUTATION — ANNUAL ABERRATION
— PROPER MOTION

41. In the preceding chapter we considered the corrections necessary to be applied to observed coördinates in order to reduce them to the center of the earth. We shall now consider the corrections which must be applied to the apparent geocentric coördinates.

While the relative positions of the fixed stars change very slowly, — and in most cases no change at all has been detected, — their *apparent* coördinates are continually varying. These variations are divided into two general classes, secular and periodic.

**Secular variations** are very slow and nearly regular changes covering long periods of time; so that for a few years, and in some cases for centuries, they may be regarded as proportional to the time.

**Periodic variations** are changes which pass quickly from one extreme value to another, so that they cannot be treated as proportional to the time except for very short intervals.

The planes of the ecliptic and equator are subject to slow motions, which give rise to variations in the obliquity of the ecliptic and in the positions of the equinoxes. The coördinates of the stars therefore undergo changes which do not arise from the motions of the stars themselves, but from a shifting of the planes of reference and the origin of coördinates. The forces producing these changes are variable, and while the variations of the coördinates are progressive, they are not uniform. They may be regarded as made up of two parts, viz.: a secular variation called **precession**, and a periodic variation called **nutation**.

Owing to annual aberration [see § 37] the stars are not seen in their true positions, but are apparently displaced toward that point of the sphere which the earth is approaching, thus giving rise to periodic variations of their apparent coördinates.

In the case of stars having **proper motions**, — that is, apparent individual motions due to motions of the stars themselves, and to the motion of the solar system in space, — their positions on the sphere change, and give rise to secular variations of the coördinates.

**42.** In order that we may define the positions of the ecliptic and equator at any instant, it will be convenient to adopt the positions of these planes at some epoch as **fixed planes**, to which their positions at any other instant may be referred. Let their positions at the beginning of the year 1800 be adopted as the mean ecliptic and equator at that instant.

The **true equator** and **ecliptic** at any instant are the *real* equator and ecliptic at that instant. Their positions are affected by precession and nutation.

The positions of the **mean equator** and **ecliptic** at any instant are the positions these circles would occupy at that instant if they were affected by precession, but not by nutation.

The **mean place** of a star at any instant is its position referred to the mean equator and ecliptic of that instant. It is affected by precession and proper motion.

The **true place** of a star is its position referred to the true equator and ecliptic. It is the mean place plus the variation due to nutation.

The **apparent place** of a star is the position in which it would be seen by an observer (at the center of the earth). It is the true place plus the variation due to annual aberration.

**43.** In solving the problems considered in the following chapters we require to know the *apparent* right ascensions and declinations of the celestial objects at the instants when they are observed. The apparent places of the sun, moon, major planets, and several hundred of the brighter stars, are given in the Ephemeris at intervals such that their places for any instant may be obtained by interpolation. But occasionally it is desirable to employ stars not included in this list. If the mean places of these stars are given in the Ephemeris for the beginning of the year* they must be reduced, by means of the proper formulæ, to the apparent places at the times of observation. If we observe stars which are not contained in the Ephemeris we must refer for their positions to the general **Star Catalogues**, which contain their mean places for the beginning of a certain year. These must be reduced to the corresponding mean places for the beginning of the year in which the observations are made, and thence to the apparent places as before. We shall now very briefly consider the matters essential to these reductions.

### PRECESSION

**44.** If from the figure of the earth we subtract a sphere whose radius is equal to the earth's polar radius, there will remain a shell of matter symmetrically situated with reference to the equator. The attractions of the sun and moon on this shell tend to draw it into coincidence with the ecliptic. This tendency is resisted by the diurnal rotation of the earth. The combined effect of these forces is to shift the plane of the equator, without changing the obliquity of the ecliptic, in such a way that its intersection with the ecliptic continually moves to the west. This causes a common annual increase in the longitudes of the

---

* This does not refer to the ordinary or tropical year, but to the fictitious year, which begins at the instant when the sun's mean longitude is 280°.

stars, which is called the **luni-solar precession**. It affects the longitudes, right ascensions and declinations, but not the latitudes.

The attractions of the other planets upon the earth tend to draw it out of the plane in which it is revolving around the sun. The effect is to shift the plane of the ecliptic in such a way that its intersection with the equator moves to the east. This causes a small annual decrease of the right ascensions of the stars, called the **planetary precession**. It affects the longitudes, latitudes and right ascensions, but not the declinations.

The attractions of the planets produce a slight change in the obliquity of the ecliptic. Its annual effect upon the coördinates of the stars is combined with the luni-solar and planetary precession, the whole being called the **general precession**.

**45.** These motions are illustrated in Fig. 9. Let $CV_0$ be the fixed or mean ecliptic at the beginning of the year 1800, $UV_0$ the mean equator, and $V_0$ the mean equinox. By the action of the sun and moon in the time $t$ the equator is shifted to the position $QV_1$, the vernal equinox moves from $V_0$ to $V_1$, and $V_0 V_1$ is the luni-solar precession in the interval $t$. By the attraction of the planets the ecliptic is shifted to the position $CV_3$, the vernal equinox moves from $V_1$ to $V$, and $V_1 V$ is the planetary precession in the interval $t$. Let

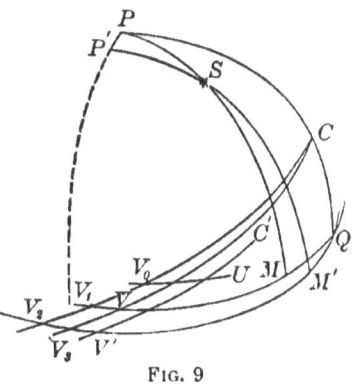

Fig. 9

$\epsilon_0$ = the mean obliquity of the ecliptic for $1800 = C V_0 U$,
$\epsilon_1$ = the obliquity of the fixed ecliptic for $1800 + t = C V_1 Q$,
$\epsilon$ = the mean obliquity of the ecliptic for $1800 + t = C V Q$,

$\psi$ = the luni-solar precession in the interval $t = V_0 V_1$,
$\vartheta$ = the planetary precession in the interval $t = V_1 V$,
$\psi_1$ = the general precession in the interval $t = CV - CV_0$.

The values of these quantities are, according to Struve and Peters, referred to the beginning of the year 1800,

$$\left.\begin{array}{l} \epsilon_0 = 23°\ 27'\ 54''.22, \\ \epsilon_1 = \epsilon_0 + 0''.00000735\ t^2, \\ \epsilon = \epsilon_0 - 0''.4738\ t - 0''.0000014\ t^2, \\ \psi = 50''.3798\ t - 0''.0001084\ t^2, \\ \vartheta = 0''.15119\ t - 0''.00024186\ t^2, \\ \psi_1 = 50''.2411\ t + 0''.0001134\ t^2. \end{array}\right\} \quad (120)$$

**46.** *Given the mean right ascension and declination $(a, \delta)$ of a star for any date $1800 + t$, required the mean right ascension and declination $(a', \delta')$ for any other date $1800 + t'$.*

In Fig. 9 let $CV_2$ be the ecliptic of 1800, $V_1 Q$ the mean equator of $1800 + t$, and $V_2 Q$ the mean equator of $1800 + t'$. If we distinguish by accents the values given by (120) for the time $t'$ we have

$$V_1 V_2 = \psi' - \psi, \quad QV_1 V_2 = 180° - \epsilon_1, \quad QV_2 V_1 = \epsilon_1'.$$

Now let

$$QV_1 = 90° - z, \quad QV_2 = 90° + z', \quad V_1 QV_2 = \theta,$$

and we have [*Chauvenet's Sph. Trig.*, § 27]

$$\cos \tfrac{1}{2} \theta \sin \tfrac{1}{2}(z' + z) = \sin \tfrac{1}{2}(\psi' - \psi) \cos \tfrac{1}{2}(\epsilon_1' + \epsilon_1),$$
$$\cos \tfrac{1}{2} \theta \cos \tfrac{1}{2}(z' + z) = \cos \tfrac{1}{2}(\psi' - \psi) \cos \tfrac{1}{2}(\epsilon_1' - \epsilon_1),$$
$$\sin \tfrac{1}{2} \theta \sin \tfrac{1}{2}(z' - z) = \cos \tfrac{1}{2}(\psi' - \psi) \sin \tfrac{1}{2}(\epsilon_1' - \epsilon_1),$$
$$\sin \tfrac{1}{2} \theta \cos \tfrac{1}{2}(z' - z) = \sin \tfrac{1}{2}(\psi' - \psi) \sin \tfrac{1}{2}(\epsilon_1' + \epsilon_1).$$

But $\tfrac{1}{2}(z' - z)$ and $\tfrac{1}{2}(\epsilon_1' - \epsilon_1)$ are very small arcs, and we can write

$$\tan \tfrac{1}{2}(z' + z) = \tan \tfrac{1}{2}(\psi' - \psi) \cos \tfrac{1}{2}(\epsilon_1' + \epsilon_1), \quad (121)$$

$$\tfrac{1}{2}(z' - z) = \frac{\tfrac{1}{2}(\epsilon_1' - \epsilon_1)}{\tan \tfrac{1}{2}(\psi' - \psi) \sin \tfrac{1}{2}(\epsilon_1' + \epsilon_1)}, \quad (122)$$

$$\sin \tfrac{1}{2} \theta = \sin \tfrac{1}{2}(\psi' - \psi) \sin \tfrac{1}{2}(\epsilon_1' + \epsilon_1), \quad (123)$$

which determine $z'$, $z$ and $\theta$ very accurately.

$V$ and $V'$ are the positions of the mean equinox for $1800 + t$ and $1800 + t'$. Representing the planetary precessions $V_1 V$ and $V_2 V'$ by $\vartheta$ and $\vartheta'$, we have

$$VQ = 90° - z - \vartheta, \qquad V'Q = 90° + z' - \vartheta';$$

and since for the star $S$ we have $a = VM$ and $a' = V'M'$, we obtain

$$MQ = 90° - z - \vartheta - a, \qquad M'Q = 90° + z' - \vartheta' - a'.$$

Then if $P$ and $P'$ are the poles of the mean equator at $1800 + t$ and $1800 + t'$, and if we put

$$A = a + \vartheta + z, \qquad A' = a' + \vartheta' - z', \qquad (124)$$

we have, in the triangle $SPP'$,

$$PS = 90° - \delta, \qquad P'S = 90° - \delta', \qquad PP' = V_1 Q V_2 = \theta,$$
$$SPP' = 90° - MQ = A, \qquad SP'P = 90° + M'Q = 180° - A'.$$

Substituting these in (2) and (3) we obtain

$$\cos \delta' \cos A' = \cos \delta \cos A \cos \theta - \sin \delta \sin \theta,$$
$$\cos \delta' \sin A' = \cos \delta \sin A,$$

from which we deduce

$$\cos \delta' \sin (A' - A) = \cos \delta \sin A \sin \theta (\tan \delta + \tan \tfrac{1}{2} \theta \cos A), \qquad (125)$$
$$\cos \delta' \cos (A' - A) = \cos \delta - \cos \delta \cos A \sin \theta (\tan \delta + \tan \tfrac{1}{2} \theta \cos A); \qquad (126)$$

or, putting

$$p = \sin \theta (\tan \delta + \tan \tfrac{1}{2} \theta \cos A), \qquad (127)$$

we have

$$\tan (A' - A) = \frac{p \sin A}{1 - p \cos A}. \qquad (128)$$

From the triangle $SPP'$ we can also obtain [*Chauvenet's Sph. Trig.*, § 22]

$$\tan \tfrac{1}{2} (\delta' - \delta) = \tan \tfrac{1}{2} \theta \frac{\cos \tfrac{1}{2} (A' + A)}{\cos \tfrac{1}{2} (A' - A)}. \qquad (129)$$

Having determined $\epsilon_1$, $\psi$, $\vartheta$, $\epsilon_1'$, $\psi'$ and $\vartheta'$ from (120), $z$, $z'$ and $\theta$ from (121), (122) and (123), and $A$ from (124), we obtain $a'$ from (127), (128) and (124), and $\delta'$ from (129).

E

*Example.* The mean place of *Polaris* for 1755.0 was

$$a = 0^h 43^m 42^s.11, \quad \delta = + 87° 59' 41''.11;$$

neglecting proper motion what will be its mean place for 1900.0?

In this case $t = -45$ and $t' = +100$, and we find, from (120),

| | | | |
|---|---|---|---|
| $\epsilon_1$ | 23° 27' 54''.23488 | $\epsilon_1'$ | 23° 27' 54''.29350 |
| $\psi$ | — 37 47 .31 | $\psi'$ | + 1 23 56 .90 |
| $\vartheta$ | — 7 .29 | $\vartheta'$ | + 12 .70 |

and therefore

| | |
|---|---|
| $\frac{1}{2}(\epsilon_1' + \epsilon_1)$ | 23° 27' 54''.26 |
| $\frac{1}{2}(\psi' - \psi)$ | + 1  0 52 .10 |
| $\frac{1}{2}(\epsilon_1' - \epsilon_1)$ | + 0 .02931 |

| | | | |
|---|---|---|---|
| $\tan \frac{1}{2}(\psi' - \psi)$ | 8.248163 | $\sin A$ | 9.3125989 |
| $\cos \frac{1}{2}(\epsilon_1' + \epsilon_1)$ | 9.962513 | $\log p$ | 9.6050849 |
| $\frac{1}{2}(z' + z)$ | 0° 55' 50''.14 | $\cos A$ | 9.9906400 |
| $\log \frac{1}{2}(\epsilon_1' - \epsilon_1)$ | 8.467016 | $\log p \cos A$ | 9.5957249 |
| $\cot \frac{1}{2}(\psi' - \psi)$ | 1.751837 | Sub* | 0.2176761 |
| $\operatorname{cosec} \frac{1}{2}(\epsilon_1' + \epsilon_1)$ | 0.399910 | $\tan(A' - A)$ | 9.1353599 |
| $\log \frac{1}{2}(z' - z)$ | 0.618763 | $A' - A$ | 7° 46' 36''.67 |
| $\frac{1}{2}(z' - z)$ | 4''.16 | $A'$ | 19 37 47 .01 |
| $z'$ | 0° 55' 54''.30 | $a' = A' + z' - \vartheta'$ | 20 33 28 .61 |
| $z$ | 0 55 45 .98 | $a'$ | $1^h 22^m 13^s.91$ |
| $\sin \frac{1}{2}(\psi' - \psi)$ | 8.248095 | | |
| $\sin \frac{1}{2}(\epsilon_1' + \epsilon_1)$ | 9.600090 | $A' + A$ | 31° 28' 57''.35 |
| $\frac{1}{2}\theta$ | 0° 24' 14''.16 | $\tan \frac{1}{2}\theta$ | 7.8481943 |
| $a$ | 10 55 31 .65 | $\cos \frac{1}{2}(A' + A)$ | 9.9833991 |
| $A = a + z + \vartheta$ | 11 51 10 .34 | $\sec \frac{1}{2}(A' - A)$ | 0.0010009 |
| $\tan \frac{1}{2}\theta$ | 7.848194 | $\tan \frac{1}{2}(\delta' - \delta)$ | 7.8325943 |
| $\cos A$ | 9.990640 | $\frac{1}{2}(\delta' - \delta)$ | 0° 23' 22''.85 |
| $\tan \frac{1}{2}\theta \cos A$ | 7.838834 | $\delta' - \delta$ | 0 46 45 .70 |
| $\tan \delta$ | 1.4557773 | $\delta'$ | 88 46 26 .81 |
| Add* | 0.0001049 | | |
| $\sin \theta$ | 8.1402027 | | |
| $\log p$ | 9.6050849 | | |

**47.** *Required the annual precession in right ascension and declination at any time* $1800 + t$.

---

* Zech's *Tafeln der Additions- und Subtractions-Logarithmen* are used here.

The precession for one year being small we can put, in (124) and (125), without sensible error,
$$\delta' = \delta, \quad \sin(A' - A) = (A' - A)\sin 1'', \quad \sin A = \sin a,$$
$$\sin \theta \tan \tfrac{1}{2}\theta = 0, \quad \sin \theta = \theta \sin 1'',$$
and obtain
$$A' - A = a' - a + (\vartheta' - \vartheta) - (z' + z) = \theta \sin a \tan \delta. \qquad (130)$$
For (121) and (123) we may write
$$z' + z = (\psi' - \psi)\cos \epsilon_1,$$
$$\theta = (\psi' - \psi)\sin \epsilon_1.$$
Substituting these in (130) and dividing by $t' - t$, we obtain
$$\frac{a' - a}{t' - t} = \frac{\psi' - \psi}{t' - t}\cos \epsilon_1 - \frac{\vartheta' - \vartheta}{t' - t} + \frac{\psi' - \psi}{t' - t}\sin \epsilon_1 \sin a \tan \delta.$$
Similarly, from (129) we can obtain
$$\frac{\delta' - \delta}{t' - t} = \frac{\psi' - \psi}{t' - t}\sin \epsilon_1 \cos a.$$
In order to express the rate of change in $a$ and $\delta$ at the instant $1800 + t$ we must let $t' - t$ become very small. Passing to the limit we have
$$\frac{da}{dt} = \frac{d\psi}{dt}\cos \epsilon_1 - \frac{d\vartheta}{dt} + \frac{d\psi}{dt}\sin \epsilon_1 \sin a \tan \delta,$$
$$\frac{d\delta}{dt} = \frac{d\psi}{dt}\sin \epsilon_1 \cos a.$$
If we let $dt$ equal one year, and put
$$m = \frac{d\psi}{dt}\cos \epsilon_1 - \frac{d\vartheta}{dt}, \quad n = \frac{d\psi}{dt}\sin \epsilon_1,$$
we obtain for the annual precession at $1800 + t$
$$\frac{da}{dt} = m + n \sin a \tan \delta, \qquad (131)$$
$$\frac{d\delta}{dt} = n \cos a. \qquad (132)$$
From (120) we find
$$\frac{d\psi}{dt}\cos \epsilon_1 = (50''.3798 - 0''.0002168\, t)\cos \epsilon_1$$
$$= 46''.2135 - 0''.0001989\, t,$$
$$\frac{d\vartheta}{dt} = \quad 0''.1512 - 0''.0004837\, t;$$

and therefore

$$m = 46''.0623 + 0''.0002849\, t, \qquad (133)$$
$$n = 20''.0607 - 0''.0000863\, t. \qquad (134)$$

Except for stars near the poles and for long intervals of time, formulæ (131) and (132) are very convenient for computing the whole precession between two dates. Thus if it is required to determine the precession in $a$ and $\delta$ from $1800 + t$ to $1800 + t,'$ we first obtain approximate values of $a$ and $\delta$ for the middle date $1800 + \frac{1}{2}(t + t')$. Using these values we then compute the annual precession for this date, which is approximately the *average* annual precession for the interval $t' - t$, and thence the whole precession by multiplying this by $t' - t$.

It is convenient to have the values of $m$ and $n$ given by (133) and (134) tabulated as follows:

| Date | $\tfrac{1}{15} m$ | log $\tfrac{1}{15} n$ | log $n$ |
|---|---|---|---|
| 1750 | 3ˢ.06987 | 0.126348 | 1.302439 |
| 1760 | 3 .07006 | 0.126330 | 1.302421 |
| 1770 | 3 .07025 | 0.126311 | 1.302402 |
| 1780 | 3 .07044 | 0.126292 | 1.302383 |
| 1790 | 3 .07063 | 0.126274 | 1.302365 |
| 1800 | 3 .07082 | 0.126255 | 1.302346 |
| 1810 | 3 .07101 | 0.126236 | 1.302327 |
| 1820 | 3 .07120 | 0.126218 | 1.302309 |
| 1830 | 3 .07139 | 0.126199 | 1.302290 |
| 1840 | 3 .07158 | 0.126180 | 1.302271 |
| 1850 | 3 .07177 | 0.126162 | 1.302253 |
| 1860 | 3 .07196 | 0.126143 | 1.302234 |
| 1870 | 3 .07215 | 0.126124 | 1.302215 |
| 1880 | 3 .07234 | 0.126106 | 1.302197 |
| 1890 | 3 .07253 | 0.126087 | 1.302178 |
| 1900 | 3 .07272 | 0.126068 | 1.302159 |
| 1910 | 3 .07291 | 0.126050 | 1.302141 |
| 1920 | 3 .07310 | 0.126031 | 1.302122 |
| 1930 | 3 .07329 | 0.126012 | 1.302103 |
| 1940 | 3 .07348 | 0.125994 | 1.302085 |

PRECESSION 53

*Example.* The mean place of $\beta$ *Orionis* for 1850.0 was

$$a = 5^h\, 7^m\, 19^s.856, \qquad \delta = -\, 8°\, 22'\, 44''.74\,;$$

neglecting proper motion, find its mean place for 1900.0.

Using the values of $m$ and $n$ for the middle date 1875.0, and $a$ and $\delta$ for 1850.0, we may obtain very nearly the annual precession in $a$ and $\delta$ for 1862.5 from (131) and (132).

| | | | |
|---|---|---|---|
| log $\tfrac{1}{15}n$ | 0.126115 | | |
| sin $a$ | 9.988429 | | |
| tan $\delta$ | 9.168186$_n$ | | |
| log | 9.282730$_n$ | | |
| number | $-\,0^s.19175$ | log $n$ | 1.302206 |
| $\tfrac{1}{15}m$ | 3.07224 | cos $a$ | 9.357543 |
| $\dfrac{da}{dt}$ | 2.88049 | $\dfrac{d\delta}{dt}$ | $4''.568$ |

The approximate coördinates of the star for 1875.0 are therefore

$$a = 5^h\, 8^m\, 31^s.87, \qquad \delta = 8°\, 20'\, 50''.5.$$

Using these values we have

| | | | |
|---|---|---|---|
| log $\tfrac{1}{15}n$ | 0.126115 | | |
| sin $a$ | 9.988955 | | |
| tan $\delta$ | 9.166514$_n$ | | |
| log | 9.281584$_n$ | | |
| number | $-\,0^s.10124$ | log $n$ | 1.302206 |
| $\tfrac{1}{15}m$ | 3.07224 | cos $a$ | 9.347705 |
| $\dfrac{da}{dt}$ | 2.88100 | $\dfrac{d\delta}{dt}$ | $4''.46592$ |

These are very nearly the exact values of the annual precession for 1875.0, and the mean place for 1900.0 is therefore

$$a' = 5^h\, 9^m\, 43^s.906, \qquad \delta' = -\, 8°\, 19'\, 1''.44,$$

which is practically identical with that given by the rigorous method of § 46.

In many star catalogues the annual precession in $a$ and $\delta$ is given for each star for the epoch of the catalogue, by means of which the approximate place of the star for the middle time is found at once, and the first approximation made above is avoided.

### PROPER MOTION

**48.** The proper motion of a star has already been defined to be an apparent motion of the star itself on the surface of the sphere. It is assumed to take place in the arc of a great circle, and to be uniform. The proper motions in right ascension and declination are the components of this motion in and perpendicular to the equator. They are variable since the equator is a moving circle, and it must be specified to which equator they refer.

When a star's place is required to be known very accurately, its position should be taken from as many catalogues as possible. In order that the data thus obtained may be properly combined, a thorough knowledge of the subject of proper motion is essential.

**49.** *Given the observed mean places* $(\alpha, \delta)$ *of a star for* $1800 + t$ *and* $(\alpha', \delta')$ *for* $1800 + t'$, *required the annual proper motion.*

Starting from the first observed place and computing the precession for the interval $t' - t$ by the methods of § 46 or § 47, let the resulting place for $1800 + t'$ be $\alpha_1, \delta_1$. The discrepancies $\alpha' - \alpha_1$ and $\delta' - \delta_1$ are due to proper motion, and the annual proper motion for the interval is

$$d\alpha' = \frac{\alpha' - \alpha_1}{t' - t}, \qquad d\delta' = \frac{\delta' - \delta_1}{t' - t}, \tag{135}$$

*referred to the equator of* $1800 + t'$.

Starting from the second observed place, computing the precession for the interval $t - t'$, and applying it to $\alpha'$ and $\delta'$, let the resulting place for $1800 + t$ be $\alpha_2, \delta_2$. The annual proper motion for the interval is

$$d\alpha = \frac{\alpha - \alpha_2}{t - t'}, \qquad d\delta = \frac{\delta - \delta_2}{t - t'}, \tag{136}$$

*referred to the equator of* $1800 + t$.

PROPER MOTION     55

*Example.* The mean places of *Polaris* for 1755.0 and 1900.0 given in *Newcomb's Standard Stars* are

for 1755.0,  $a = 0^h 43^m 42^s.11$,  $\delta = + 87° 59' 41''.11$,
for 1900.0,  $a' = 1\ 22\ 33.76$,  $\delta' = + 88\ 46\ 26\ .66$;

determine the proper motion referred to the equator of 1900.0.

By applying the precession to the place for 1755.0 the place for 1900.0 was found to be, § 46,

$$a_1 = 1^h 22^m 13^s.91, \quad \delta_1 = + 88° 46' 26''.81,$$

and therefore, by (135), the annual proper motion of *Polaris* referred to the equator of 1900.0 is

$$da' = + 0^s.1369, \quad d\delta' = - 0''.00103.$$

**50.** *Given the proper motion* $(da, d\delta)$ *referred to the equator of* $1800 + t$, *required the corresponding proper motion* $(da', d\delta')$ *referred to the equator of* $1800 + t'$, *and vice versa.*

When the star $S$ (Fig. 9) moves on the surface of the sphere it causes variations in all the parts of the triangle $SP'P$, except $P'P$. The solution of the present problem requires a knowledge of the relations existing between these variations.

If in a spherical triangle ABC we suppose all the parts except a to vary, we can write [*Chauvenet's Sph. Trig.*, § 153, (286) and (287)]

$$\sin c\, dB = \sin A\, db - \sin a \cos A \sin B \operatorname{cosec} A\, dC,$$
$$dc = \cos A\, db + \sin a \sin B\, dC$$

Substituting in these, from § 46,

$$a = PP' = \theta, \quad db = d(SP) = - d\delta, \quad dc = d(SP') = - d\delta',$$
$$dB = d(SP'P) = d(180° - A') = - da', \quad dC = d(SPP') = dA = da,$$

and putting $\gamma$ for A, we obtain

$$\cos \delta'\, da' = \sin \gamma\, d\delta + \cos \delta \cos \gamma\, da, \tag{137}$$
$$d\delta' = \cos \gamma\, d\delta - \cos \delta \sin \gamma\, da, \tag{138}$$

in which $\gamma$ is determined by

$$\sin \gamma = \sin \theta \sin A \sec \delta' = \sin \theta \sin A' \sec \delta, \qquad (139)$$
$$\cos \gamma = (\cos \theta - \sin \delta \sin \delta') \sec \delta \sec \delta'. \qquad (140)$$

These determine the proper motion for $1800 + t'$ in terms of that for $1800 + t$.

From (137) and (138) we obtain

$$\cos \delta \, da = \cos \delta' \cos \gamma \, da' - \sin \gamma \, d\delta', \qquad (141)$$
$$d\delta = \cos \delta' \sin \gamma \, da' + \cos \gamma \, d\delta', \qquad (142)$$

which determine the proper motion for $1800 + t$ in terms of that for $1800 + t'$.

*Example.* The proper motion of *Polaris* referred to the equator of 1900.0 is

$$da' = + 0^s.1369 = + 2''.0535, \qquad d\delta' = - 0''.00103.$$

Deduce the proper motion referred to the equator of 1755.0.

Using $\theta$ and $A$ from § 46 and $\delta' = + 88° 46' 26''.66$ we find, from (139) and (140),

$$\sin \gamma = 9.131493, \qquad \cos \gamma = 9.995984,$$

and therefore from (141) and (142) we obtain

$$da = + 1''.2480 = + 0^s.0832, \qquad d\delta = + 0''.00493.$$

**51.** *Given the proper motion $(da, d\delta)$ and the mean place $(a, \delta)$ of a star for the epoch $1800 + t$, required its mean place $(a', \delta')$ for the epoch $1800 + t'$.*

The proper motion for the whole interval $t' - t$ is first computed and applied to the mean place for $1800 + t$. With the resulting values of $a$ and $\delta$, which we shall denote by $a_1$ and $\delta_1$, the precession is computed and applied to $a_1$ and $\delta_1$. The result is the star's mean place for $1800 + t'$.

If the proper motion $(da', d\delta')$ is given for the epoch $1800 + t'$, we first compute the precession, using $a$ and $\delta$, and then apply the proper motion for the interval.

*Example 1.* Given the mean place and proper motion of *Polaris* for 1755.0,

$$a = 0^h 13^m 42^s.11, \quad \delta = + 87° 59' 41''.11,$$
$$da = + 0^s.0832, \quad d\delta = + 0''.00493,$$

required the mean place for 1900.0.
The proper motion for the interval is

$$+ 0^s.0832 \times 145 = + 12^s.06, \quad + 0''.00493 \times 145 = + 0''.71.$$

Therefore
$$a_1 = 0^h 43^m 54^s.17, \quad \delta_1 = + 87° 59' 41''.82.$$

Employing these values in the example of § 46 we find for 1900.0

$$a' = 1^h 22^m 33^s.76, \quad \delta' = + 88° 46' 26''.66.$$

*Example 2.* The proper motion of $\beta$ *Orionis* referred to the equator of 1900.0 is

$$da = - 0^s.00027, \quad d\delta = - 0''.0061.$$

Include this in the example of § 47.
The proper motion for the interval is

$$- 0^s.00027 \times 50 = - 0^s.013, \quad - 0''.0061 \times 50 = - 0''.30;$$

and therefore the mean place for 1900.0 is

$$a' = 5^h 9^m 43^s.893, \quad \delta' = - 8° 19' 1''.74.$$

**52.** It will be seen from § 47 that the annual precession is a slowly varying quantity. The change in its value in one hundred years is called the **secular variation** of the precession. Many star catalogues give not only the mean place and annual precession of a star but also the secular variation and proper motion. In this case the reduction of the mean place of a star from the epoch of the catalogue $1800 + t$ to that for $1800 + t'$ is readily made. For if

$p = $ the annual precession for the epoch $1800 + t$,
$\Delta p = $ the secular variation,
$\mu = $ the proper motion,

the reduction for the interval $t' - t$ will be the annual change for the middle time multiplied by $t' - t$, or

$$\left[p + \frac{\Delta p}{100} \cdot \frac{t'-t}{2} + \mu\right](t' - t). \tag{143}$$

which form applies both to the right ascension and the declination.

*Example.* *Newcomb's Standard Stars* gives the following data for $\beta$ *Orionis* for the epoch 1850.0:

$a = 5^h 7^m 19^s.856,$   $\delta = -8° 22' 44''.74,$
$p = + 2^s.87999,$   $p' = + 4''.5687,$
$\Delta p = + 0^s.00400,$   $\Delta p' = - 0''.4109,$
$\mu = - 0^s.00025,$   $\mu' = - 0''.0061;$

required the mean place for 1900.0.

Substituting these values in (143) we obtain the reductions for the interval,

for $a$, $+ 144^s.038$,   for $\delta$, $+ 222''.99$,

and the mean place for 1900.0 is, therefore,

$a' = 5^h 9^m 43^s.894,$   $\delta' = - 8° 19' 1''.75.$

53. Many of the problems of practical astronomy require that the star places be determined with the utmost accuracy. In such cases the observed coördinates of a star — along with the corresponding epochs of observation — are taken from as many star catalogues as possible, and combined by the method of least squares in order to determine the most probable values of the star's coördinates and proper motion at any given time.

Suppose the star's right ascension is given in $n$ catalogues for the epochs of observation $t_1, t_2, \cdots t_n$, and that the most probable values of the right ascension and proper motion are required for the epoch $t$. Apply the precession up to the instant $t$ to each of the catalogue positions, and let the results be $a_1, a_2, \cdots a_n$. Let $\mu$ be the star's proper motion referred to the equator of the epoch $t$, and let $a$ be its right ascension at that instant. Then we shall have $n$ equations

$$\left.\begin{aligned} a &= a_1 + \mu\,(t - t_1), \\ a &= a_2 + \mu\,(t - t_2), \\ &\ldots\quad\ldots\quad\ldots\quad\ldots \\ a &= a_n + \mu\,(t - t_n), \end{aligned}\right\} \quad (144)$$

from which to determine the most probable values of $\mu$ and $a$. To put them in a form suitable for solution, let $\mu_0$ and $a_0$ be approximate values of $\mu$ and $a$, and let $\Delta\mu$ and $\Delta a$ be small corrections to them, so that

$$\mu = \mu_0 + \Delta\mu, \quad (145)$$
$$a = a_0 + \Delta a; \quad (146)$$

then equations (144) take the form

$$\left.\begin{aligned}(t - t_1)\,\Delta\mu - \Delta a + [a_1 + (t - t_1)\mu_0 - a_0] &= 0, \\ \ldots\quad\ldots\quad\ldots\quad\ldots\quad\ldots\quad\ldots \\ (t - t_n)\,\Delta\mu - \Delta a + [a_n + (t - t_n)\mu_0 - a_0] &= 0.\end{aligned}\right\} \quad (147)$$

The solution of these equations will determine the most probable values of $\Delta\mu$ and $\Delta a$, and therefore of $\mu$ and $a$, provided the $n$ original data are of equal weight. If they are of unequal weight, as will almost always be the case, the equations (147) must be multiplied by the proper factors before proceeding to their solution.

The weights to be assigned to the data from different star catalogues depend upon many factors. The instruments and methods of observation and reduction employed, the skill of the observers, and the number of individual observations upon which the printed results depend, must all be taken into account. Familiarity with the methods of meridian circle work and considerable experience with star catalogues are necessary acquirements for assigning suitable weights. Tables of relative weights in the introductions to *Newcomb's Standard Stars* and *Boss's 500 Stars* will serve as partial guides.

*Example.* It is required to determine the most probable values of the right ascension and the proper motion in right ascension of $\delta$ *Trianguli*, for the epoch 1900.0.

Observations of this star are contained in twenty or more well-known catalogues. We shall select ten of them, as below. The first column contains the name of the catalogue, the second the epoch of observation, the third the catalogue right ascensions corrected for the precession up to 1900.0, and the fourth the relative weights. Except for small errors of observation, the discrepancies in column

| Catalogue | Epoch of Obs'n | $a$ | W't | $a$ 1900.0 |
|---|---|---|---|---|
| Auwers-Bradley | 1755.0 | $2^h\ 10^m\ 43^s.72$ | 2 | $2^h\ 10^m\ 57^s.03$ |
| Lalande | 1794.9 | 47.03 | 1 | 56.69 |
| Piazzi | 1812.9 | 48.42 | 1 | 56.43 |
| Abo | 1830.0 | 50.43 | 2 | 56.86 |
| Edinburgh | 1842.9 | 51.68 | 2 | 56.93 |
| Pulkowa | 1855.0 | 52.72 | 4 | 56.86 |
| Greenwich N. 7 yr. | 1864.0 | 53.63 | 4 | 56.94 |
| " 9 yr. | 1872.0 | 54.36 | 4 | 56.93 |
| " 10 yr. | 1880.0 | 55.07 | 4 | 56.91 |
| Cincinnati | 1890.6 | 55.95 | 4 | 56.82 |

three are due to proper motion. Comparing the first and last observations, we find for an approximate value of the annual proper motion, $\mu_0 = +0^s.09$; and therefore for an approximate value of the right ascension at 1900.0, $a_0 = 2^h\ 10^m\ 56^s.80$. We may now write equations (147), thus:

$$\left.\begin{array}{r}145.0\,\Delta\mu - \Delta a - 0.05 = 0, \\ \cdots\cdots\cdots\cdots \\ 9.4\,\Delta\mu - \Delta a \pm 0.00 = 0.\end{array}\right\} \quad (148)$$

Multiplying these equations by the square roots of their respective weights, and combining them, we obtain

REDUCTION TO APPARENT PLACE 61

and thence
$$+ 94792 \Delta\mu - 1283 \Delta a - 82.07 = 0,* \\ - 1283 \Delta\mu + 28 \Delta a + 0.29 = 0, \quad (149)$$

$$\Delta\mu = + 0^s.0019, \qquad \Delta a = + 0^s.077.$$

Substituting these and $\mu_0$ and $a_0$ in (145) and (146), we obtain the most probable proper motion and right ascension for 1900.0,

$$\mu = + 0^s.09 + 0^s.0019 = + 0^s.0919,$$
$$a = 2^h 10^m 56^s.80 + 0^s.077 = 2\ 10^m 56^s.877.$$

The last column of the table above contains the individual right ascensions corrected for proper motion. The modern observations are in good agreement.

An entirely analogous method would be used to determine the most probable values of the declination, and the proper motion in declination.

### REDUCTION TO APPARENT PLACE

**54.** The mean place of a star for the beginning of the required year having been obtained by any of the above methods, it remains to determine its apparent place for any given instant. Thus if we desire the apparent place for a time $\tau$ from the beginning of the year, we obtain the mean place by adding the precession and proper motion for the interval $\tau$, then the true place by adding the nutation, and finally the apparent place by adding the annual aberration. The reduction to the mean place could be performed as before; we could determine the nutation by evaluating the long and tedious nutation formulæ, the deduction of which belongs to physical astronomy; and we could obtain the annual aberration from equations deduced by methods analogous to those of § 38. But this process is laborious, and the general equations are never used except in a few highly specialized problems.

---

* Another computer may not exactly reproduce these coefficients, on account of neglected decimals.

By judiciously combining the terms of the various formulæ involved in the reductions, Bessel was able to propose two simple and closely related methods, which are now in common use. We shall consider them in the following section.

**55.** *Given the mean place* $(a, \delta)$ *of a star for the beginning of the year, required its apparent place* $(a', \delta')$ *for any instant* $\tau$.*

(*a*) The reduction is made by the formulæ

$$a' = a + \tau\mu + Aa + Bb + Cc + Dd + \tfrac{1}{15} E, \qquad (150)$$
$$\delta' = \delta + \tau\mu' + Aa' + Bb' + Cc' + Dd', \qquad (151)$$

in which $\tau\mu$ and $\tau\mu'$ represent the proper motion and $Aa$ and $Aa'$ the precession in the interval $\tau$; $Bb + \tfrac{1}{15} E$ and $Bb'$ the nutation, and $Cc + Dd$ and $Cc' + Dd'$ the annual aberration at the instant $\tau$. $A$, $B$, $C$, $D$, and $E$ are the **Besselian star-numbers.** They are functions of the time. The American Ephemeris gives their general values on p. 280, and tabulates the logarithms of $A$, $B$, $C$, and $D$ for every day of the year on pp. 281–284. The value of $E$ is given in the same place. It is a slowly varying quantity whose value never exceeds $0''.05$, and it can generally be neglected.

$a$, $b$, $c$, $d$, $a'$, $b'$, $c'$, and $d'$ are **Bessel's star-constants.** They are functions of the star's place and the obliquity of the ecliptic, and are defined by the equations

$$\left.\begin{array}{ll} a = \tfrac{1}{15} m + \tfrac{1}{15} n \sin a \tan \delta, & a' = n \cos a, \\ b = \tfrac{1}{15} \cos a \tan \delta, & b' = -\sin a, \\ c = \tfrac{1}{15} \cos a \sec \delta, & c' = \tan \epsilon \cos \delta - \sin a \sin \delta, \\ d = \tfrac{1}{15} \sin a \sec \delta, & d' = \cos a \sin \delta. \end{array}\right\} \quad (152)$$

In some star catalogues the logarithms of the star-constants are given for each star. But these values become obsolete in a few years, and must be computed anew from (152), since $m$, $n$, $a$, $\delta$, and $\epsilon$ are variable quantities.

---

* See § 43, footnote, and the American Ephemeris, p. 280.

*Example.* Required the apparent place of 38 *Lyncis* for the upper transit at Ann Arbor, 1891 March 16.

From the *Berliner Jahrbuch*, p. 180, star 135, we find for 1891.0,

$\alpha = 9^h\,12^m\,3^s.071,$   $\delta = +37° 15' 48''.43,$
$\mu = -0.0030,$   $\mu' = -0.114.$

The upper transit occurs therefore at Washington sidereal time $9^h\,39^m$, or $1^h\,53^m$ before mean midnight. Taking the values of $\log A$, etc., from the American Ephemeris, p. 281, for this instant, and the values of $\log a$, etc., from the *Jahrbuch*, p. 329, star 135, the computation is conveniently arranged as below.

| | | | |
|---|---|---|---|
| $\log a$ 0.5744 | $\log b$ 8.5764$_n$ | $\log c$ 8.7943$_n$ | $\log d$ 8.7485 |
| $\log A$ 9.0221$_n$ | $\log B$ 0.5804$_n$ | $\log C$ 1.2722$_n$ | $\log D$ 0.1239 |
| $\log a'$ 1.1734$_n$ | $\log b'$ 9.8254$_n$ | $\log c'$ 8.7763$_n$ | $\log d'$ 9.6533$_n$ |

| | | | | |
|---|---|---|---|---|
| $\alpha$ | $9^h\,12^m\,3^s.671$ | $\delta$ | | $+37° 15' 48''.43$ |
| $\tau\mu$ | $-0.001$ | $\tau\mu'$ | | $-0.02$ |
| $Aa$ | $-0.395$ | $Aa'$ | | $+1.57$ |
| $Bb$ | $+0.143$ | $Bb'$ | | $+2.55$ |
| $Cc$ | $+1.165$ | $Cc'$ | | $+1.12$ |
| $Dd$ | $+0.075$ | $Dd'$ | | $-0.60$ |
| $\tfrac{1}{15}E$ | $-0.003$ | | | |
| $\alpha'$ | 9  12  4.655 | $\delta'$ | | $+37\ 15\ 53.05$ |

This method of reduction should be employed when the star-constants are given in the catalogues with sufficient accuracy, or when the apparent places of the same star are required for several dates.

In using the data of reduction furnished by the British and French annuals and catalogues, the computer must be careful to follow their formulæ; for while the *form* of reduction usually agrees with the American and German form, the *notation* is different, $A$ and $B$ in the one corresponding to $C$ and $D$ respectively in the other. This applies also to the American Ephemeris previous to 1865.

(*b*) When the catalogues do not give the values of $\log a$, $\log b$, etc., and when only one or two places of the same star are desired, another form of reduction is preferable. If we put

$$f = \tfrac{1}{15} m A + \tfrac{1}{15} E, \qquad h \sin H = C,$$
$$g \sin G = B, \qquad\qquad h \cos H = D,$$
$$g \cos G = n A, \qquad\qquad i = C \tan \epsilon,$$

the formulæ (150) and (151) become

$$a' = a + \tau\mu + f + \tfrac{1}{15} g \sin(G + a) \tan \delta + \tfrac{1}{15} h \sin(H + a) \sec \delta, \quad (153)$$
$$\delta' = \delta + \tau\mu' + g \cos(G + a) + h \cos(H + a) \sin \delta + i \cos \delta, \quad (154)$$

in which the terms involving $f$, $g$ and $G$ denote the precession and nutation, and the terms involving $h$, $H$ and $i$, the annual aberration. These auxiliary quantities are called the **independent star-numbers**. The values of $\tau, f$, $G$, $H$, $\log g$, $\log h$ and $\log i$ are given in the American Ephemeris, pp. 285–292, for every day of the year.

*Example.* Required the apparent place of 38 *Lyncis* for the upper transit at Ann Arbor, 1891 March 16.

Using the data given above, the computation is conveniently made as below.

| | | | | | |
|---|---|---|---|---|---|
| $G$ | 240° 59′ | $\log g$ | 0.6387 | $a$ | 9ʰ 12ᵐ 3ˢ .671 |
| $a$ | 138  1 | $\cos(G + a)$ | 9.9757 | $\tau\mu$ | − 0 .001 |
| $H$ | 274  3 | | | $f$ | − 0 .326 |
| | | $\log h$ | 1.2733 | (1) | + 0 .072 |
| $\log \tfrac{1}{15}$ | 8.8239 | $\cos(H + a)$ | 9.7887 | (2) | + 1 .240 |
| $\log g$ | 0.6387 | $\sin \delta$ | 9.7821 | $a'$ | 9 12 4 .656 |
| $\sin(G + a)$ | 9.5126 | $\log(4)$ | 0.8441 | | |
| $\tan \delta$ | 9.8813 | | | $\delta$ | + 37° 15′ 48″.43 |
| $\log(1)$ | 8.8565 | $\log i$ | 0.9096ₙ | $\tau\mu'$ | − 0 .02 |
| | | $\cos \delta$ | 9.9008 | (3) | + 4 .12 |
| $\log \tfrac{1}{15}$ | 8.8239 | | | (4) | + 6 .98 |
| $\log h$ | 1.2733 | | | (5) | − 6 .46 |
| $\sin(H + a)$ | 9.8969 | | | $\delta'$ | + 37 15 53 .05 |
| $\sec \delta$ | 0.0992 | | | | |
| $\log(2)$ | 0.0933 | | | | |

# CHAPTER V

### ANGLE AND TIME MEASUREMENT

**56.** The degree of refinement to which an observer can carry the determination of his geographical position and the time depends in general upon the accuracy attainable in *pointing the telescope*, in *reading the angle* corresponding to the pointing, and in *noting the time* when the pointing is made. In general, these elements are of equal importance. For any given telescope the first depends upon the observer's skill. In the last two the observer's skill is assisted by various mechanical devices.

#### THE VERNIER

**57.** An angle is usually measured by means of a **graduated circle**, or arc, whose center is at the vertex of the angle. Closely fitting upon the graduated arc of the circle and centered with it is another graduated arc called the **vernier**, which is so arranged upon an arm that it moves with reference to the circle when the telescope moves. The angle to be read is that included between the zero line of the circle and the zero line of the vernier. The zero of the vernier generally falls between two consecutive lines on the circle. The angle corresponding to the whole divisions can be read off at once; it is the object of the vernier to determine the fractional part of a division. It is so constructed that $n$ of its divisions are equal in length to $n - 1$ divisions of the circle. If we let

$d$ = the value of one division of the circle,
$d'$ = the value of one division of the vernier,

we have
$$(n-1)d = nd',$$
or
$$d - d' = \frac{d}{n}. \tag{155}$$

$d - d'$ is called the **least reading** of the vernier. If now the zero of the vernier coincides with a division line of the circle, the circle reading gives the required angle at once. If the first vernier line coincides with a circle line, the zero of the vernier is $d - d'$ beyond a line of the circle, and the circle reading must be increased by the least reading. If the second vernier line is in coincidence with a circle line, the circle reading must be increased by twice the least reading, etc. For example, the value of a division of a sextant is 10', and 60 divisions of the vernier correspond in length to 59 divisions of the circle. The least reading is $10' \div 60 = 10''$. In measuring a certain angle the zero of the vernier fell between 42° 40' and 42° 50', and the 26th line of the vernier coincided with a circle line. The required reading was 42° 40' + 26 × 10'' = 42° 44' 20''. In practice no computation is necessary, the number of minutes being read directly from the numbers on the vernier.

In the accompanying illustration, Fig. 10, there are two verniers on the upper graduated arc: one to the right of

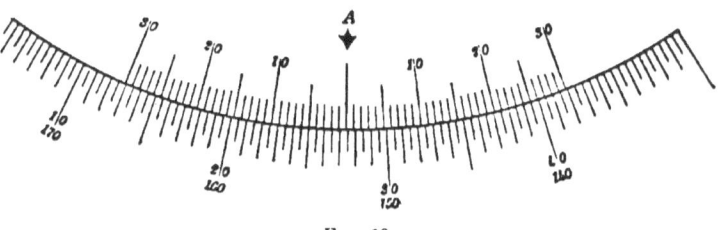

Fig. 10

A, to be used in connection with the inner set of readings on the circle numbered from 10° to the right; and one to the left of A, to be used in connection with the outer set of readings numbered from 140° to the left. From (155) it follows that the least reading of the vernier is 1'. The circle reading, when read to the right, is 27° 25'; and when read to the left, 152° 35'.

## THE READING MICROSCOPE

**58.** In very fine instruments the vernier is replaced by a **reading microscope,** the optical axis of which is perpendicular to the plane of the graduated circle. The microscope is so adjusted that an image of the circle divisions is formed in the common focus of the objective and ocular. In the same focus are two very fine **micrometer wires** (usually spider-lines) which either intersect at a small angle, or are parallel and close together. In the former case they are adjusted so that the bisector of their acute angle is parallel to the image of the circle graduation seen nearest the middle of the field of view. In the latter case, the wires are made parallel to that graduation. They are stretched upon a light frame whose plane is parallel to the plane of the circle, and which may be moved in that plane in the direction at right angles to the visible graduations by turning a fine **micrometer screw.** Fixed upon the projecting end of the screw is a cylindrical **micrometer head.** This is graduated into either 60 or 100 parts, and is used for reading the fractional parts of a revolution of the screw, the readings being made with reference to a fixed index. The whole number of revolutions is indicated by a scale sometimes inside, and again outside, of the microscope.

Let the micrometer screw be turned until the wires are in the center of the field of view and the reading of the head is zero. The position now occupied by the wires is the fixed point of reference. The angle to be read is that included between the zero of the circle and this point. If now the micrometer wires coincide with a line of the circle the desired reading is obtained at once. If they fall beyond a certain line the fractional part of a division is determined by moving the wires from the point of reference into coincidence with the line. The distance passed over is determined from the micrometer reading and the known angular value of one revolution of the screw. In setting the wires upon any circle division the last motion of the micrometer should always take place in the direction which increases the tension on the screw, any lost or dead motion being thereby avoided.

When the microscope is properly adjusted, a whole number of revolutions of the screw corresponds exactly to the distance between two consecutive circle lines. But this adjustment once made does not remain, owing to changes of temperature, etc. It is customary to determine from time to time the error which arises and allow for it. This is called the **error of runs**. Its value is found by measuring several divisions in different parts of the circle, and taking the mean of the measures in order to eliminate as far as possible any errors in the graduations. To illustrate, let a circle be graduated to $5'$, let the value of a revolution of the screw be $1'$, and let the head be divided into 60 parts. Let the mean of the measures of ten divisions in different parts of the circle be 4 revolutions and 56.4 divisions of the head. The correction for runs *per minute of arc* is $+ 3''.6 \div 5 = + 0''.72$. Let an angle be read such that the circle graduation employed is $62° 15'$, and the micrometer reading is 2 rev. 15.9 div. The correction for runs is $+ 1''.6$, and the angle is, therefore, $62° 17' 17''.5$.

Better still, the program of observations should if possible include microscope readings on the two graduations nearest the middle of the field of view, instead of on only one. Each complete observation would then furnish the necessary data to correct for error of runs. This excellent practice is illustrated in the example of § 126.

## ECCENTRICITY

**59.** The center of the arm which carries the vernier or microscope never coincides exactly with the center of the circle, and an error due to this eccentricity enters into the circle readings. In Fig. 11, let $C'$ be the center of the vernier, $C$ the center of the circle, $D$ the point of intersection of the circle $DAB$ and the line $CC'$ produced, $A$ the zero point of the circle, and $M$ the position of the vernier or microscope. The pointing of the telescope corresponds to the direction $C'M$ while the circle reading refers to the direction $CM$. The correction for eccentricity is therefore $C'MC$. To find its value let

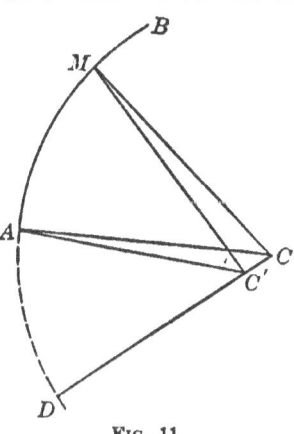

Fig. 11

$CC'$ = the eccentricity = $e$,
$\epsilon$ = the correction for eccentricity = $C'MC$,
$M$ = the observed reading of the circle,
$R$ = the true reading of the circle = $M + \epsilon$,
$r$ = the radius of the circle,
$\eta$ = the angle $DC'A$,
$\nu$ = the angle $AC'M$.

From the triangle $C'MC$ we can write

$$r \sin \epsilon = e \sin (\nu + \eta).$$

Since $r$ is the unit radius and $\epsilon$ is very small, we have

$$\epsilon = \frac{e}{\sin 1''} \sin (\nu + \eta) = e'' \sin (\nu + \eta), \qquad (156)$$

and the true reading of the circle is

$$R = M + e'' \sin (\nu + \eta). \qquad (157)$$

The vernier arm $MC'$ is often produced to the opposite point of the circle, which call $M'$, and carries another vernier or microscope. The minutes and seconds of the circle reading at this point being obtained (a second vernier is not necessary to determine the degrees), if $M'$ is the observed reading, the true reading $R$ is given by

$$R = M' + e'' \sin(180° + \nu + \eta) = M' - e'' \sin(\nu + \eta).$$

Combining this with (157) we have

$$R = \tfrac{1}{2}(M + M'); \qquad (158)$$

from which it appears that the eccentricity is fully eliminated by taking the mean of two readings 180° apart. It can be shown also that it is eliminated by using the mean reading of any number of equidistant microscopes.

### THE MICROMETER

**60.** If the angle between two points is smaller than the angular diameter of the field of the telescope, it is most easily and accurately measured by means of a micrometer. This is the same in principle as that used on the reading microscope, save that the **movable wire** is usually composed of a single thread, and is generally accompanied by other wires. There is usually one **fixed wire** parallel to the movable wire, and often at least one **transverse fixed wire** perpendicular to it. The arrangement of the wires varies to meet the requirements of different problems and the preferences of the observers. The plane of the wires is in the common focal plane of the objective and eyepiece of the telescope.

The micrometer is so constructed that it can be, and always is, rotated about the line of sight until the micrometer wire is perpendicular to the plane of the angle which it is desired to measure. In the transit instrument the movable wire of the micrometer is vertical, and in the zenith telescope it is horizontal. [See § 110, and Fig. 20.] The modern meridian circle is provided with both horizontal and vertical micrometer wires. The **filar microm-**

eter of an equatorial telescope is arranged so that the wires can be turned into any position, and their direction may be determined by means of a graduated position circle.

Figure 12 represents the filar micrometer of the 12-inch equatorial of the Lick Observatory, by Alvan Clark & Sons. The wires are in the box $S$. To render them visible at night, they are illuminated by the lamp $L$, supported by the framework $JQRPV$ (designed by Burnham), and counterbalanced by $IHF$. The graduated position circle $XY$ remains fixed with reference to the telescope, whereas the micrometer box (and the illuminating apparatus) may be rotated about the line of sight so as to place the wires in any direction. Their direction will be indicated by the circle readings at $X$ and $Y$. Further, by turning the screw $EE'$ the micrometer box, with the entire system of wires, can be moved in a direction parallel to the micrometer screw. The system of wires may therefore be given motions

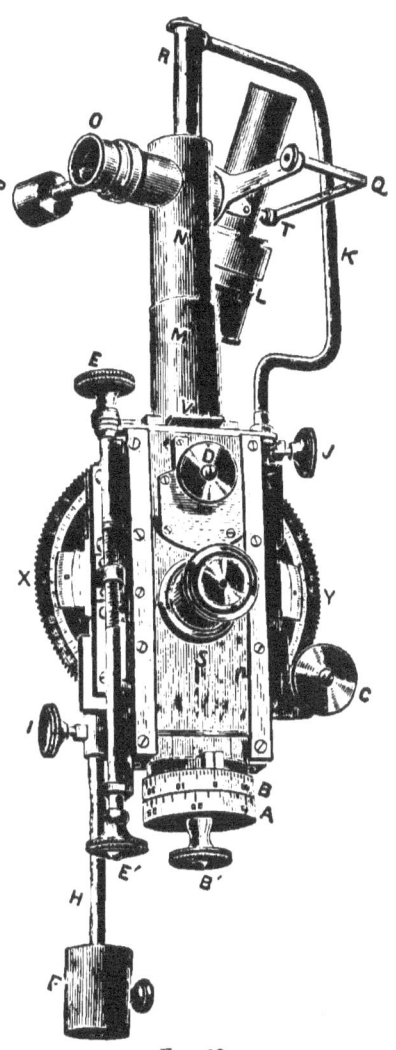

FIG. 12

of rotation and translation, to place them in any desired position.

The light from the lamp shines in the direction $TO$, but at the intersection of the tubes $TO$ and $NM$ there is a diagonal mirror which reflects the light in the direction $NV$ into the box. The mirror may be rotated by rotating a disk at $O$, thereby varying the intensity of the illumination. If electric lighting is available, the oil lamp $L$ should be replaced by a small incandescent lamp, as the latter has many practical advantages.

The distance between the fixed and the movable wires, or the distance between two positions of the movable wire, is indicated by the readings of the graduated micrometer heads $A$ and $B$: $A$ indicating the whole number of revolutions of the screw and $B$ the fractional parts of a revolution. To convert the readings into arc, the value of one revolution of the screw must be known.

**61.** *The* **angular value of a revolution** *of the micrometer screw* depends upon the pitch of the screw and the focal length of the telescope. It may be found in several ways.

(*a*) By the methods described above, measure with the micrometer any known angle, and divide the number of seconds in the angle by the corresponding number of revolutions of the screw.

If the distance between two stars is measured, the true distance must be corrected for refraction. A method commonly employed consists in measuring the difference of declination of two selected stars in the Pleiades when that group is near the meridian. The positions of these stars are very accurately known, pairs of almost any desired distance can be selected, and the correction for refraction is simple.

*Example.* The difference of declination of *d Ursæ Majoris* and *Groombridge* 1564 was measured with the movable wire of the micrometer of the zenith telescope of the Detroit Observatory, when on the meridian, 1891 March 28. Barom. 29.206 inches, Att. Therm. 58°.0 F., Ext. Therm. 37°.5 F. Find the value of a revolution of the screw. The zenith level was read immediately after bisecting each star, in order to correct for any change in the pointing of the telescope. The value of one division of the level is 2".74.

| Star | α | Apparent δ | Level | | Micrometer |
|---|---|---|---|---|---|
| | | | n | s | |
| d Ursæ Maj. | 9ʰ 25ᵐ | + 70° 18' 46".5 | 40.4 | 18.7 | 48.566 |
| Gr. 1564. | 9 33 | + 69 44 12 .7 | 41.3 | 19.5 | 2.598 |

The correction for level (see § 62) is $0.85\,d = 2''.3$, by which amount the measured distance must be increased, or the difference of the declinations decreased. The difference of the refractions for the two stars is $0''.7$, by which amount the difference of the declinations must be decreased. The corrected difference of the declinations is $34'\,30''.8$. Therefore the value of one revolution of the screw is $45''.05$.

(b) A more accurate value is obtained by observations on one of the close circumpolar stars. The telescope is directed so that the star is just entering the field, and will be carried through the center by its diurnal motion. The micrometer is revolved so that the micrometer wire will be perpendicular to the diurnal motion of the star when it passes through the center of the field. The wire is set just in advance of the star, the time of transit of the star over it is noted, and the micrometer is read. The wire is moved forward one revolution, or a part of a revolution, and the transit observed as before. In this way the observations are carried nearly across the field.

In Fig. 13, let $P$ be the pole, $EP$ the observer's meridian, $SS'$ the diurnal path of a star, $AS$ the position of the micrometer wire when at the center of the field and coincident with an hour circle $PM$, and $BS'$ (parallel to $AS$) any other position of the wire. Now let $m_0$ be the micrometer reading, $t_0$ the hour angle, and $T_0$ the sidereal time when the star is at $S$, and let $m$, $t$ and $T$ be the corresponding quantities when the star is at $S'$, and let $R$ be the value of one revolution of the screw. Through $S'$ pass an arc of a great circle $S'C$ perpendicular to $AS$. Then in the triangle $CS'P$, right-angled at $C$, we have

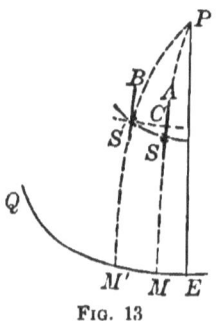

Fig. 13

$$CS' = (m - m_0)R, \quad S'P = 90° - \delta, \quad CPS' = t - t_0 = T - T_0;$$

and we can write

$$\sin[(m - m_0)R] = \sin(T - T_0)\cos\delta;$$

or, since $(m - m_0)R$ is always a small angle,

$$(m - m_0)R = \sin(T - T_0)\frac{\cos\delta}{\sin 1''}. \tag{159}$$

Similarly, for another observation,

$$(m' - m_0)R = \sin(T_1' - T_0)\frac{\cos\delta}{\sin 1''}.$$

Combining these to eliminate the zero point,

$$(m' - m)R = \sin(T' - T_0)\frac{\cos\delta}{\sin 1''} - \sin(T - T_0)\frac{\cos\delta}{\sin 1''}, \tag{160}$$

from which the value of $R$ is obtained. The micrometer readings are supposed to increase with the time.

The times of transit are supposed to be noted by means of a sidereal time-piece. If its rate [§ 64] is large it must be allowed for. If a mean time-piece is used the intervals $T - T_0$ must be converted into sidereal intervals.

The resulting value of $R$ is slightly in error on account of refraction, since the star is observed at unequal zenith distances. But the effect of refraction is inappreciable if the observations are made near the meridian. The method is therefore advantageous for a meridian instrument with a micrometer in right ascension, the star being observed at upper or lower culmination. However, any variations in the azimuth or level constants of the instrument during the progress of the observations introduce errors in the results. If $a$ and $b$ are the values of these constants at the beginning of the series of transits and $a'$ and $b'$ their values at the close of the series, it can be shown from the theory of the transit instrument [Chapter VII], that the distance between the first and last positions of the wires has been decreased by the quantity

$$(a' - a) \sin(\phi \mp \delta) + (b' - b) \cos(\phi \mp \delta), \qquad (161)$$

which divided by the corresponding difference of the micrometer readings is the correction to the value of one revolution of the screw. The lower signs are for lower culmination.

The azimuth constants are determined by observing suitable pairs of stars before and after the series of micrometer transits is taken, according to the methods described later. The level constants are determined by the method of § 62. The variations of azimuth and level may be considered to be uniform and proportional to the time. For a meridian instrument properly mounted the variation of the azimuth may be neglected without a sacrifice of accuracy.

*Example.* *Polaris* was observed at lower culmination at Ann Arbor, 1891 March 28, to determine the value of a revolution of the micrometer screw of the transit instrument. The micrometer was set at every three-tenths of a revolution, and one hundred and fifty transits observed. The times were noted by means of a sidereal chronometer which was $16^m 30^s.6$ slow. The position of *Polaris* was

$$\alpha = 1^h 17^m 46^s.0, \qquad \delta = + 88° 43' 40''.25,$$

American Ephemeris, p. 304; and therefore the chronometer time of lower culmination was $T_0 = 18^h 1^m 15^s.4$. A few of the observations and their reduction are given below. Each printed observation is the mean of three consecutive original observations.

| $m$ | $T$ | $T - T_0$ | $T - T_0$ | $\sin(T - T_0)$ | $(m - m_0)R$ |
|---|---|---|---|---|---|
| 7.8 | 12$^h$ 23$^m$ 12$^s$.3 | − 38$^m$ 3$^s$.1 | − 9° 30′ 46″.5 | 9.218194$_n$ | − 756″.83 |
| 8.7 | 25 17 .3 | 35 58 .1 | 8 59 31 .5 | 9.193953$_n$ | 715 .75 |
| 9.6 | 27 19 .7 | 33 55 .7 | 8 28 55 .5 | 9.168793$_n$ | 675 .46 |
| 10.5 | 29 22 .0 | 31 53 .4 | 7 58 21 .0 | 9.142070$_n$ | 635 .15 |
| 11.4 | 31 22 .0 | 29 53 .4 | 7 28 21 .0 | 9.114111$_n$ | 595 .55 |
| 20.4 | 51 42 .3 | 9 33 .1 | 2 23 16 .5 | 8.619771$_n$ | 190 .80 |
| 21.3 | 53 43 .3 | 7 32 .1 | 1 53 1 .5 | 8.516822$_n$ | 150 .53 |
| 22.2 | 55 46 .0 | 5 29 .4 | 1 22 21 .0 | 8.379348$_n$ | 109 .69 |
| 23.1 | 57 47 .0 | 3 28 .4 | 0 52 6 .0 | 8.180547$_n$ | 69 .40 |
| 24.0 | 59 50 .0 | − 1 25 .4 | − 0 21 21 .0 | 7.793121$_n$ | − 28 .44 |
| 25.8 | 13 3 54 .0 | + 2 38 .6 | + 0 39 39 .0 | 8.061960 | + 52 .82 |
| 26.7 | 5 58 .3 | 4 42 .9 | 1 10 43 .5 | 8.313268 | 94 .21 |
| 27.6 | 7 59 .5 | 6 44 .1 | 1 41 1 .5 | 8.468092 | 134 .55 |
| 28.5 | 10 0 .0 | 8 44 .6 | 2 11 9 .0 | 8.581389 | 174 .66 |
| 29.4 | 12 3 .7 | 10 48 3 | 4 42 4 .5 | 8.673281 | 215 .82 |
| 38.4 | 32 24 .3 | 31 8 .9 | 7 47 13 .5 | 9.131914 | 620 .48 |
| 39.3 | 34 28 .3 | 33 12 .9 | 8 18 13 .5 | 9.159630 | 661 .20 |
| 40.2 | 36 32 .2 | 35 16 .8 | 8 49 12 .0 | 9.185629 | 702 .16 |
| 41.1 | 38 34 .0 | 37 18 .6 | 9 19 39 .0 | 9.209722 | 742 .21 |
| 42.0 | 13 40 37 .0 | + 39 21 .6 | + 9 50 24 .0 | 9.232735 | + 782 .60 |

Subtracting the 1st from the 11th, the 2d from the 12th, etc., we have

| $m' - m$ | $(m' - m)R$ | $R$ | $v$ | $v^2$ |
|---|---|---|---|---|
| 18.0 | 809″.65 | 44″.981 | − 0″.065 | 0.0042 |
| 18.0 | 809 .96 | 44 .998 | − 0 .048 | 0.0023 |
| 18.0 | 810 .01 | 45 .001 | − 0 .045 | 0.0020 |
| 18.0 | 809 .81 | 44 .989 | − 0 .057 | 0.0032 |
| 18.0 | 811 .37 | 45 .076 | + 0 .030 | 0.0009 |
| 18.0 | 811 .28 | 45 .071 | + 0 .025 | 0.0006 |
| 18.0 | 811 .73 | 45 .096 | + 0 .050 | 0.0025 |
| 18.0 | 811 .85 | 45 .103 | + 0 .057 | 0.0032 |
| 18.0 | 811 .61 | 45 .089 | + 0 .043 | 0.0018 |
| 18.0 | 811 .04 | 45 .058 | + 0 .012 | 0.0001 |

$$R = 45.046 \qquad \Sigma v^2 = 0.0208$$

$$\text{Probable error}^* = \pm 0.674 \sqrt{\frac{0.0208}{10 \times 9}} = \pm 0''.010.$$

* See Appendix C, § 1.

By reducing the whole series of transits the value of $R$ and its probable error were found to be

$$R = 45''.059 \pm 0''.006.$$

From the level readings $b = +5''.17$, $b' = +7''.05$; and from observations for azimuth on $\beta$ *Cassiopeiæ* and 4 *H. Draconis*, and on $\theta$ *Bootis* and 36 *H. Cassiopeiæ*, $a = -9''.15$, $a' = -8''.79$. Substituting these in (161) and dividing by 46, the difference of the first and last micrometer readings of the original series, we have as a correction to $R$, $-0''.017$, and therefore

$$R = 45''.042 \pm 0''.006.$$

There is an indication from the individual results for $R$ that its value increases as the micrometer readings increase. This irregularity should be fully investigated by further observations, and allowed for in refined observations if it proves to be real.

The value of a revolution is affected by changes of temperature. To determine the rate of change, observations should be made on several nights at widely different temperatures. If $R$ is the value of a revolution at the temperature $\tau$, $R_0$ the value at the temperature $50°$, and $x$ the correction to $R_0$ for a rise of $1°$ in temperature, each night's observations furnish an equation of the form

$$R = R_0 + (\tau - 50°)x. \quad (162)$$

The solution of these equations by the method of least squares gives the most probable values of $R_0$ and $x$, and therefore of $R$.

(c) If the micrometer is designed for the measurement of zenith distances, the micrometer wire being horizontal, the observations are made at the time of the star's **greatest western** or **eastern elongation**. This occurs when the vertical circle of the star is tangent to its diurnal circle. At this time the micrometer wire is parallel to the star's hour circle. If $m_0$, $t_0$ and $T_0$ refer to the instant of greatest elongation and $m$, $t$ and $T$ to any other instant, the formula (159) is applicable to this case. At the instant of greatest elongation the parallactic angle $ZOP$, Fig. 1, is 90° for western and 270° for eastern elongation, and we can write

$$\cos t_0 = \tan \phi \cot \delta, \quad \cos z_0 = \sin \phi \operatorname{cosec} \delta, \quad T_0 = a + t_0. \quad (163)$$

Set the telescope at the zenith distance $z_0$ when the star is just entering the instrument. Note the time of transit over the micrometer wire; and, as before, carry the observations nearly across the field. Any change in the zenith distance of the telescope during the progress of the observations will affect the resulting value of $R$. The amount of the change will be indicated by the zenith level and can be allowed for. The level should be read after each transit is observed. If $l_0$ is the level reading, *i.e.* the reading of the level scale for the middle of the bubble, at the time $T_0$, $l$ the level reading at the time $T$, and $d$ the value in arc of a division of the level, we have

$$(m - m_0) R = \pm \sin (T - T_0) \frac{\cos \delta}{\sin 1''} + (l - l_0) d; \quad (164)$$

and for another observation

$$(m' - m_0) R = \pm \sin (T' - T_0) \frac{\cos \delta}{\sin 1''} + (l' - l_0) d.$$

Whence

$$(m' - m) R = \pm \sin (T' - T_0) \frac{\cos \delta}{\sin 1''} \mp \sin (T - T_0) \frac{\cos \delta}{\sin 1''} + (l' - l) d, \quad (165)$$

in which the lower sign is for eastern elongation. The micrometer readings are supposed to increase with the time for western elongations, and the level readings to increase towards the north.

The resulting value of $R$ must be corrected for refraction. From the values of $z_0$ and $R$, the zenith distances corresponding to the first and last observations can be obtained, and thence the refractions. The difference of the refractions divided by the difference of the first and last micrometer readings is the amount by which the value of $R$ must be decreased.

If both $R$ and $d$ are unknown a close approximation to the value of $R$ is obtained by neglecting the term $(l' - l)d$. With this value of $R$ the value of $d$ is computed (§ 63) and substituted in (165), and the corrected value of $R$ obtained. A second approximation to the value of $d$ will rarely be required.

## THE LEVEL

**62. The spirit level** consists of a sealed glass tube, ground on the upper interior surface to the arc of a circle of large radius, and nearly filled with alcohol or ether. The bubble of air occupying the space not filled by the liquid is always at the highest point of the curve. Therefore a change in the relative elevations of the ends of the tube causes a motion of the bubble, the amount of which is read from a scale marked on the surface of the glass. The level is adapted to the determination of the angle which a nearly horizontal line makes with the horizon, or the very small angle moved over by a telescope.

The level tube is mounted and attached to astronomical instruments in various ways, but there is one general method of using it. Let the divisions of the scale be numbered in both directions from zero at the center, and let $d$ be the angular value of one division. If the level be placed on a truly horizontal line — say, for convenience, an east and west line — the center of the bubble will not be at zero, owing to the non-adjustment of the level. If the center is $x$ divisions from the zero, the error of the level is $dx$. Let the level be placed on a line inclined to the horizon at an angle $b$, and let the reading of the west end of the bubble be $w$ and the east end $e$. Then the elevation of the west end of the line is given by

$$b = \tfrac{1}{2}(w - e)d \mp dx.$$

Now let the level be reversed in direction and let the reading of the west end be $w'$ and the east end $e'$. Then

$$b = \tfrac{1}{2}(w' - e')d \pm dx.$$

Combining these values of $b$ we have

$$b = \tfrac{1}{4}[(w + w') - (e + e')]d; \qquad (166)$$

from which it appears that the error of the level is eliminated by reversing. A positive value of $b$ will indicate that the west end of the line is higher than the east end.

Whenever it is possible the level should be read several times, the same number of readings being made in each position, — level *direct* and level *reversed*, — care being taken to remove the level from its bearings after each reading is made.

*Example.* The inclination of the axis of a transit instrument is required from the following level readings, the value of one division being $1''.88$.

|          | $w$  | $e$  |           |       |
|----------|------|------|-----------|-------|
| Direct   | 14.1 | 9.7  | $\Sigma w$ | 53.4  |
| Reversed | 12.6 | 11.1 | $\Sigma e$ | 41.8  |
| Reversed | 12.7 | 11.1 | *8)+      | 11.6  |
| Direct   | 14.0 | 9.9  | +         | 1.45  |
| Sum      | 53.4 | 41.8 |           |       |

The axis makes an angle $+ 1.45\, d = + 2''.73$ with the horizon, the west end being higher than the east.

In case the zero of the scale is at one end of the tube and the numbers increase continuously to the other, — which is a better system, — we can show that

$$b = \tfrac{1}{4}[(w + e) - (w' + e')]d, \qquad (167)$$

in which the readings $w$ and $e$ for level direct correspond to that position of the level for which the readings increase toward the west.

---

\* It must be noticed that there are two complete observations for determining $b$, and hence the divisor is 8 instead of 4.

*Example.* Find the inclination of the axis of a transit instrument from the following level readings, the value of one division being $2''.743$.

|     | Direct |     | Reversed |          |
|-----|--------|-----|----------|----------|
| $w$ | 35.4   | $w'$ | 16.3    | 95.3     |
| $e$ | 12.4   | $e'$ | 39.4    | 111.6    |
| $w$ | 35.3   | $w'$ | 16.4    | 8) −16.3 |
| $e$ | 12.2   | $e'$ | 39.5    | − 2.037  |
| Sum | 95.3   | Sum | 111.6   |          |

The axis makes an angle $-2.037\,d = -5''.59$ with the horizon, the west end being lower than the east.

**63.** The **value of one division** of the level is determined best by means of a **level-trier**. This consists of a horizontal bar supported at one end by two bearings and at the other by a vertical micrometer screw. The level is placed on the bar and the readings of the micrometer and bubble are noted. The screw is now turned and the bubble moves to a new position. The readings of the micrometer and bubble are again noted. The angle moved over by the bar is known from the length of the bar, the pitch of the screw and the difference of the micrometer readings; whence the angular value of one division of the level may be obtained. If possible, the determination of the value of a division should be made *after* the level tube is fixed in its final mounting, rather than before.

The essential principles of the level-trier are well illustrated by Fig. 14.

FIG. 14

In the absence of a level-trier, an accurate determination of the value of a division can be obtained by means of any telescope provided with a micrometer in zenith distance. To illustrate, let an equatorial be directed upon a distant terrestrial mark directly north or south of it, and adjust the micrometer wire to parallelism with the horizon. Mount the level upon the telescope so that the vertical plane passing through the axis of the level tube is parallel to the line of sight, and so that the bubble is at one end of the scale. The mark is bisected by the micrometer wire, and the level and micrometer readings noted. The instrument is then turned through an angle such that the bubble moves to the other end of the scale. The mark is again bisected by the wire, and the level and micrometer readings noted as before. The difference of the level readings corresponds to the difference of the micrometer readings, whence the value of one division of the level can be obtained from the known value of a revolution of the micrometer screw. In general, such observations are best made on an overcast day.

*Example.* The following observations were made February 19, 1891, to determine the value of a division of the striding level of the Detroit Observatory transit instrument, the telescope being directed to a distant mark. The value of one revolution of the screw is $45''.042$. Find the value of a division of the level.

| Level | | Micrometer | Differences | | $d$ |
|---|---|---|---|---|---|
| $n$ | $s$ | | Level | Micrometer | |
| 20.9 | 1.1 | 17.019 | 18.9 | 0.772 | $0.0408\,R$ |
| 2.0 | 20.0 | 17.791 | | | |
| 1.8 | 20.2 | 17.773 | 18.8 | 0.804 | $0.0428\,R$ |
| 20.6 | 1.4 | 16.969 | | | |

The mean of eighteen observations gave $d = 0.0417\,R \pm 0.0004\,R = 1''.878 \pm 0''.018$.

The level tube should be thoroughly tested for irregularity of curvature before using. If different portions of a level give sensibly different values for a division of the scale, it should not be used in refined observations.

The value of a division should also be determined at two or more very different temperatures in order that a temperature correction may be introduced if necessary.

A level should be adjusted by the vertical adjusting screws so that the bubble will stand near the center of the tube when the level is placed on a horizontal line. It should be adjusted by the horizontal screws so that the axis of the tube will be parallel to the line whose inclination is to be measured. This adjustment is tested by revolving the level slightly about its bearings. If the readings are different when the level is equally displaced in opposite directions from the vertical plane through its bearings, the adjustment is not perfect.

### THE CHRONOMETER

**64.** A **chronometer** is a large and carefully constructed watch which is "compensated" so that changes of temperature have very little effect on the time in which the balance-wheel vibrates. It is a very accurate time-piece when properly handled, comparing favorably with the astronomical clock, and being portable is adapted to field work and navigation.

The **chronometer correction** is the amount which must be added to the reading of the chronometer face to obtain the correct time. It is + when the chronometer is slow. The **chronometer rate** is the daily increase of the chronometer correction. It is + when the chronometer is losing. It is not necessary that the correction and rate be small, though it is convenient to have the rate less than $\pm 5^s$ a day. The test of a good time-piece lies in the *uniformity* of its rate. The correction is generally allowed to increase indefinitely.

The chronometer correction is obtained by observations on the celestial objects, or by comparison with a time-piece whose correction is known. If

$\Delta T_0$ = the chronometer correction at a time $T_0$,
$\Delta T$ = the chronometer correction at a time $T$,
$\delta T$ = the chronometer rate,

we determine the rate per unit of time by

$$\delta T = \frac{\Delta T - \Delta T_0}{T - T_0}. \tag{168}$$

Conversely, if the rate and the correction at the instant $T_0$ are known, the correction at the instant $T$ is given by

$$\Delta T = \Delta T_0 + \delta T (T - T_0). \tag{169}$$

*Example.* The correction to chronometer T. S. & J. D. Negus, no. 721, was $+ 16^m 19^s.5$ at Ann Arbor mean time 1891 March $25^d 11^h$, and $+ 16^m 55^s.6$ at 1891 April $4^d 11^h$. Find the daily rate, and the correction at 1891 March $28^d 13^h$.

From (168) we find the daily rate $\delta T = + 3^s.61$. Substituting this and $T =$ March $28^d 13^h$ in (169) we find

$$\Delta T = + 16^m 19^s.5 + 3^s.61 \times 3.08 = + 16^m 30^s.6.$$

The above equations are true only when the rate is constant for the interval $T - T_0$. Such constancy can be assumed for an interval of a few days in the case of the best chronometers; but when great accuracy is required the interval between observations for determining chronometer correction should be as small as possible.

When several chronometers are employed, the correction to one is obtained by observation; and to the others, by comparison with the first. If two chronometers which keep the same kind of time are compared, it will generally happen that they do not beat together. The fraction of a second by which one beats later than the other can be estimated after some practice to within $0^s.1$ or $0^s.2$, so that the correction can be obtained to that degree of accuracy by this method.

When a sidereal chronometer is compared with a mean

time chronometer the degree of accuracy is higher. If the chronometers tick half seconds, the beats of the two will coincide once in every $183^s$, since in this interval sidereal time gains $0^s.5$ on mean solar. The ear is capable of estimating the coincidence of the beats within $0^s.02$ or $0^s.03$. When the coincidence occurs the observer notes the times indicated by the two chronometers. The correction to the one being known, a satisfactory value of the correction to the other is readily obtained.

When a chronograph [§ 68] is at hand and the chronometers are provided with break-circuits (or make-circuits), the comparisons are most conveniently and accurately made by placing the two chronometers in the chronograph circuit. The beats are recorded on the chronograph sheet and the distance between them can be measured very accurately by means of a scale.

It is convenient to use a sidereal chronometer when making observations on the stars, planets, comets, etc., and a mean time chronometer when making observations on the sun.

**65.** The observer should be able to "carry the beat" of the chronometer; that is, mentally to count the successive seconds from the tick of the chronometer without looking at it. An experienced observer will carry the beat for several minutes, estimate the times of transits of a star over several wires (or other similar phenomena) to tenths of seconds, and write them on a slip of paper without taking his eye from the telescope: then, still carrying the beat, he will *look at the chronometer face to verify his count*. This is called the "**eye and ear method**" of observing, and it is very important that every observer should be able to employ it with accuracy and perfect ease.

**66.** To obtain the best results from a chronometer the following precepts should be rigidly observed:

(*a*) It should be wound at regular intervals. If it requires winding daily it should always be wound at the same hour of the day; otherwise an unused part of the spring is brought into action and a change of rate results.

(*b*) The hands should not be moved forward oftener than is necessary, and they should not be moved backward.

(*c*) A chronometer on shipboard should be allowed to swing freely in its gimbals, so that it may always take a horizontal position; but when carried about on land it should be clamped so as to avoid the violent oscillations due to the sudden motions it receives.

(*d*) It should be kept in a dry place; as nearly at a uniform temperature as possible; away from magnetic influences; and when at rest should always be in the same position with respect to the points of the compass.

(*e*) All quick motions should be avoided: in particular, it should never be rotated rapidly about its vertical axis.

(*f*) In out-of-door use it should be protected from the direct rays of the sun.

**67.** The **astronomical clock** is a finely constructed clock whose pendulum is compensated for changes of temperature. Its rate is in general more uniform than that of a chronometer. It is one of the *fixed* instruments of an observatory, and to that extent the remarks concerning the chronometer are applicable to it.

### THE CHRONOGRAPH

**68.** The **chronograph** is a mechanical device for recording the instant when an observation is made. A sheet of paper on which the record is to be made is wrapped around a metallic cylinder which is caused to rotate once per minute by means of clock-work. A pen is attached to the armature of an electro-magnet in such a way as to press its point on the moving paper. The magnet is carried slowly along the cylinder by a screw, so that the pen traces a continuous spiral on the paper. The electro-magnet is placed in an

Fig. 15

electric circuit which passes through the chronometer or clock (or, better, through a relay connected with the timepiece), in such a way that the circuit is broken for an instant at the beginning of every second, or every other second. At each of these instants the electro-magnet releases the armature carrying the pen, the pen moves laterally for the moment, and in this way the spiral is graduated by notches to seconds of time. One notch is usually omitted at the beginning of each minute, to assist in identifying the seconds. One of the circuit wires passes through a signal-key held in the observer's hand. When a star, for example, is being observed he presses the key at the exact instant when the star is crossing a wire, thus breaking the circuit and making the record on the chronograph sheet. The beats of the chronometer being recorded on the sheet, the chronometer time when the key was pressed can be read from the sheet by means of a scale with great accuracy, and at the observer's leisure.

When the chronograph is first set in motion the observer records in his note-book the hour, minute, and second corresponding to a certain marked notch on the sheet, which serves as a reference point in identifying all the notches on the sheet.

In some forms of the chronograph the circuit is *made* by pressing the key, but the *break-circuit* is preferable.

The chronographic method is generally preferable to the eye and ear method because it relieves the mind from carrying the beat and making the record, thus allowing greater care to be given to other parts of the observation, and because more observations can be made in a given time. But in the case of transit observations of slowly-moving stars, or of very faint objects, and in many forms of micrometer observations, the eye and ear method is at least as satisfactory as the chronographic method.

A very common form of chronograph is illustrated in Fig. 15.

# CHAPTER VI

## THE SEXTANT

**69.** The sextant is an instrument especially adapted to the determination of time, latitude and longitude when extreme accuracy is not required, as in navigation and exploration. It consists essentially of a brass frame $ADC$, Fig. 16, bearing a graduated arc $AC$, a telescope $EF$, whose line of sight is parallel to the plane of the graduated arc, and the mirrors $H$ and $D$, whose planes are perpendicular to the plane of the arc. The mirror $D$, called the **index-glass**, is fixed to the **index-arm** $DB$, which revolves about $D$ at the center of the arc, and which carries a vernier at $B$. The mirror $H$, called the **horizon-glass**, is attached to the frame. The lower half of it is silvered, the upper half is left clear.

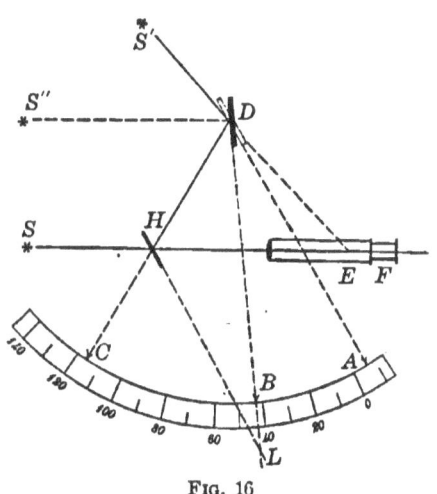

FIG. 16

Figure 17 illustrates a form of sextant commonly employed. The special parts already described in connection with Fig. 16 will be recognized without difficulty. The telescope is mounted on an adjustable standard so that its distance from the frame of the sextant may be varied by turning the screw-head at the lower end of the standard. Colored or neutral-tint glasses are mounted in front of the index and horizon glasses. They can be rotated into the paths of the sun's rays to protect the eyes while observing that body. A dense neutral-tint glass may also be screwed

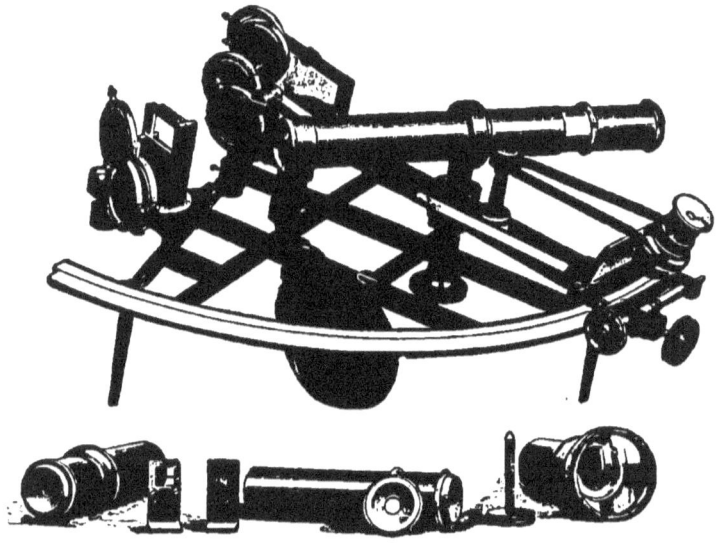

FIG. 17

over the eyepiece for the same purpose. The telescope may be replaced by others of different magnifying power, or by one with a larger object-glass for observing stars, — shown in the foreground of the cut. The index-arm carrying the vernier is furnished with a clamp and slow motion for setting accurately to any desired reading.

**70.** To illustrate the method of using a sextant and the principles involved, let it be required to measure the angle $SES'$ between the stars $S$ and $S'$. The instrument is held in the hand and the telescope directed to the star $S$. The ray $SE$ passes through the unsilvered part of $H$ and forms a *direct* image of the star at the focus $F$. The sextant is revolved about the line of sight until its plane passes through the other star $S'$. The index-arm is then moved until the *reflected* image of $S'$ is brought into the field and nearly in coincidence with the direct image of $S$. The index-arm is clamped and the two images brought into perfect coincidence by turning the slow-motion or tangent screw. If the instrument is perfectly constructed and adjusted, the required angle is given at once by the circle reading. The ray of light $S'D$, which forms the reflected image at $F$, traverses the path $S'D$-$DH$-$HF$, being reflected by the two mirrors $D$ and $H$. When the direct and reflected images coincide, the angle between the stars is twice the angle between the mirrors. That is, $SES' = 2\,HLD$, since the angle

$$SES' = 180° - EDH - EHD$$
$$= 180° - 2\,HDL - 2\,(LHD - 90°)$$
$$= 2\,(180° - HDL - LHD)$$
$$= 2\,HLD.$$

If $A$ is the position of the zero of the vernier when the two mirrors are parallel, and $B$ its position when the two images coincide, we have

$$SES' = 2\,HLD = 2\,ADB = 2\,AB. \tag{170}$$

It thus appears that to enable us to read the required angle directly from the circle, the circle reading must be twice the corresponding arc. Thus, the 120° line is really only 60° from the 0° line [or a sextant, hence the name].

An improved form of the sextant is known as the Pistor and Martins (Berlin) **prismatic sextant**, in which the horizon glass is replaced by a totally reflecting prism, occupying a somewhat different position on the frame of the instrument. Among other advantages of the prismatic form it can be used for measuring angles up to 180° and even greater; whereas with the common form, the angle is limited to about 140°.

Again, the graduated arc is sometimes a complete circle, in which case the index arm is extended over a diameter of the circle and carries a vernier on each extremity. Such an instrument is called a **reflecting circle** or **prismatic circle** according as a horizon-glass or prism is used. Its chief advantage lies in the fact that the eccentricity is eliminated by the use of two verniers 180° apart.

71. In order to obtain good results with the sextant, the instrument must be accurately adjusted, and the telescope focused; the direct and reflected images should be about equally bright; and several complete observations should be made, the mean of all being used.

The images are made equally bright by moving the telescope from or toward the frame, so as to utilize more or less of the light passing through the transparent part of the horizon-glass, or by placing colored-glass shades in front of the index-glass.

In measuring the angular distance between two stars, the images of the stars are brought into exact coincidence in the middle of the field of view. In measuring the distance of the moon from a star, the star is brought into coincidence with that point of the moon's bright limb which lies in the great circle joining the star and the center of the moon. The measured distance is then increased or decreased by the moon's semidiameter [§§ 33, 34, 35]. In the case of the sun and moon the images of the nearest limbs are made to coincide, and the measured distance is increased by the semidiameters of both objects, as before. Results obtained in this way, when corrected for any instrumental errors, are the apparent distances between the objects.

The sextant is also used for measuring the apparent altitudes of the heavenly bodies. At sea the telescope is directed to that point of the horizon which is below the object. The reflected image is brought into contact with the horizon line. When the instrument is vibrated slightly about the line of sight the image should describe a curve tangent to the horizon. The sextant reading corrected for instrumental errors and the dip of the horizon [§ 32] is the apparent altitude. If the object is the sun, the lower or upper limb is made tangent to the horizon; if the moon, the bright limb; and the sextant readings must be further corrected for semidiameter.

For observing altitudes on land an artificial horizon is used. This is a shallow basin of mercury over which is placed a roof, made of two plates of glass set at right angles to each other in a frame, to protect the mercury from agitation by air currents. The mercury forms a very perfect horizontal mirror which reflects the rays of light from the star. If the observer places his eye at some point in a reflected ray, he will see an image of the star in the mercury, whose angle of depression below the horizon is equal to the altitude of the star above the horizon. If then he directs the telescope to the image in the mercury, and brings the two images into coincidence as before, the sextant reading corrected for instrumental errors is double the apparent altitude of the star. The sun's altitude is measured by making the two images tangent externally. The corrected sextant reading is double the altitude of the lower or upper limb, according as the nearest or farthest limbs of the sun and its image in the mercury are observed.

The double altitudes of stars near the meridian are changing slowly, and the images are brought into contact by means of the slow-motion screw as before. But the double altitudes of stars at a distance from the meridian are changing rapidly, and another method is used. To illustrate, suppose the sun is observed for time when it is east of the meridian, and the altitude therefore increasing.

The upper limb is observed first. The two images are brought into the field and the index moved forward until the sextant reading is from 10′ to 20′ greater than the double altitude of the upper limb, and the instrument is clamped. The images are now slightly separated, but they are approaching. When they become tangent, the observer notes the time on the chronometer and reads the circle. The index is again moved forward from 10′ to 20′ and the contact observed as before. In this way, four or five observations are made. The double diameter of the sun is about 64′, and for observing the lower limb the index is quickly moved backward about 45′. The two images now overlap, but they are separating, and the time is noted when they become tangent. Moving the index forward as before, four or five observations are made on the lower limb.

If the sun is observed west of the meridian, the altitudes of the lower limb should be measured first.

**72.** The faces of the glass in the horizon roof should be perfectly parallel. If they are prismatic the observed altitudes are erroneous. The error is eliminated by observing one-half of a set of altitudes with the roof in one position and the other half with the roof in the reversed position, and taking the mean of all. Likewise, the glass screens in front of the index and horizon glasses must have parallel faces.

The surface of the mercury can be freed from impurities by adding a little tin-foil. The amalgam which forms can be drawn to one side of the basin by means of a card, leaving a perfectly bright surface.

### ADJUSTMENTS OF THE SEXTANT

**73.** (*a*) *The index-glass.* Place the sextant on a table, unscrew the telescope and set it in a vertical position on the graduated arc. Place the eye near the index-glass and move the index-arm toward the telescope until the telescope and its image in the mirror are seen very nearly in coincidence. Their corresponding outlines will be parallel if the index-glass is perpendicular to the plane of the arc. If they are not parallel, the glass is removed and one of the points against which it rests is filed down the proper amount. The axis of the telescope is here assumed to be perpendicular to the plane of the end on which it rests. This can be tested by rotating the telescope about its axis and noticing whether the angle between the tube and its image varies. The telescope should be set at the mean of the two positions which give the maximum and minimum values of this angle.*

(b) *The horizon-glass.* The index-glass having been adjusted, the telescope is directed to a star and the index-arm is brought near the zero of the arc. If the horizon-glass is parallel to the index-glass the reflected image will pass through the direct image when the index-arm is moved slowly to and fro. If it passes on either side of the direct image the horizon-glass needs adjustment. This is done by turning the screws provided for the purpose.

(c) *The telescope.* Two parallel wires are placed in the telescope tube. These are made parallel to the plane of the sextant by revolving the tube containing them. The line of sight is the line joining a point midway between these wires and the center of the object glass. This should be parallel to the plane of the sextant. To test the adjustment, select two well-defined objects about 120° apart, and bring the two images into coincidence on one of the side wires, and then move the sextant so as to bring the images on the other wire. If the images still coincide, the line of sight needs no adjustment. If the images are separated, the collar which holds the telescope is shifted by means of screws until the adjustment is satisfactory.

---

* This method was proposed by Professor J. M. Schaeberle: *The Sidereal Messenger*, May, 1888.

## CORRECTIONS TO SEXTANT READINGS

**74. *The index correction.*** It is seen from (170) that all angles measured with the sextant are reckoned from $A$, the point where the zero of the vernier falls when the two mirrors are parallel; whereas the circle readings are measured from 0°. The **index correction** is the reading 0° (or 360°) minus the reading at $A$. Let it be represented by $I$. The value of $I$ can be reduced to zero by rotating slightly the horizon-glass by means of screws provided for that purpose. But this adjustment is very liable to derangement, and it is customary to determine $I$ every time the sextant is used and apply it to all the sextant readings.

(*a*) To determine $I$ for correcting stellar observations, point the telescope to a star and bring the direct and reflected images into coincidence. Let the sextant reading be $R$. The index correction is given by

$$I = 0° - R. \tag{171}$$

*Example.* Determine $I$ from the following readings:

| 359° 56′ 0″ | 359° 55′ 50″ |
|---|---|
| 56 10 | 56 10 |
| 55 50 | 55 55 |

The mean of the six readings is 359° 55′ 59″.2, and therefore

$$I = 360° - 359° \ 55' \ 59''.2 = + 4' \ 0''.8.$$

(*b*) For reducing solar observations, point the telescope to the sun and bring the direct and reflected images externally tangent to each other and read the circle. Then move the reflected image over the direct image until they are again externally tangent, and read the circle. Let the readings in the two positions be $R_1$ and $R_2$, $R_1$ being the greater. The reading when the two images coincide is $\frac{1}{2}(R_1 + R_2)$, and the index correction is given by

$$I = 360° - \tfrac{1}{2}(R_1 + R_2). \tag{172}$$

CORRECTIONS TO SEXTANT READINGS 97

The observed semidiameter of the sun is given by

$$S = \tfrac{1}{4}(R_1 - R_2). \qquad (173)$$

To eliminate the effect of refraction the horizontal semidiameter should be measured.

*Example.* Find $I$ and $S$ from the following readings on the sun made Thursday, 1891 April 23.

|  | 360° 28' 35" | 359° 24' 50" |
|---|---|---|
|  | 40 | 40 |
|  | 45 | 45 |
|  | 45 | 45 |
|  | 40 | 50 |
| Means | 360 28 41 .0 | 359 24 46 .0 |

$I = 360° - 359° 56' 43''.5 = + 3' 16''.5.$

$S = \tfrac{1}{4}(1° 3' 55''.0) = 15' 58''.7.$

From the American Ephemeris, p. 56, $S = 15' 56''.3$.

**75.** *Correction for eccentricity.* The arc of a sextant being short, the eccentricity cannot be eliminated by means of two verniers 180° apart, and it must be investigated. This can be done by comparing several angles measured with the sextant with their known values obtained in some other way. Thus in Fig. 11 the sextant reading is twice the arc $AM$. The true value of the angle is obtained by correcting the reading at $M$ for eccentricity, and correcting the position of $A$ for eccentricity and index error. The true reading at $M$ is given by (157). The true reading at the zero point $A$ is given by

$$R_0 = - I + e'' \sin \eta.$$

The true value of the angle is

$$R - R_0 = M + e'' \sin(\nu + \eta) - e'' \sin \eta + I. \qquad (174)$$

But $R - R_0$ is the known value of the angle; let $d$ represent it. $M$ is the observed value of the angle; let $d'$ rep-

H

resent it. Now, since an arc on the sextant is one-half the corresponding reading, we have

$$\tfrac{1}{2}(d - d') = e'' \sin(\nu + \eta) - e'' \sin \eta + \tfrac{1}{2} I,$$

which reduces to

$$d - d' = 4 e'' \cos(\tfrac{1}{2}\nu + \eta) \sin \tfrac{1}{2}\nu + I \qquad (175)$$
$$= 4 e'' \cos \eta \sin \tfrac{1}{2}\nu \cos \tfrac{1}{2}\nu - 4 e'' \sin \eta \sin^2 \tfrac{1}{2}\nu + I.$$

If we put

$$4 e'' \cos \eta = x, \qquad 4 e'' \sin \eta = y, \qquad (176)$$

we have

$$\sin \tfrac{1}{2}\nu \cos \tfrac{1}{2}\nu \, x - \sin^2 \tfrac{1}{2}\nu \, y + I = d - d'. \qquad (177)$$

This equation involves three unknown quantities, $x$, $y$, $I$. Three measured angles, each furnishing an equation of the form (177), are required for the solution of the problem.

There are several ways in which to obtain the value of $d - d'$ at any point of the arc.

(a) For those who have access to a meridian circle, the most direct process known is the ingenious method proposed by Professor Schaeberle in *der Astronomische Nachrichten*, no. 2832.

(b) When the latitude of the observer and the time are accurately known, make a series of measures of the double altitudes of a star just before and after its meridian passage. The observed double altitude at the instant of transit is obtained from these measures by the method of § 87. The apparent double altitude at the instant is obtained at once from the known declination, latitude and refraction. The latter minus the former is $d - d'$.

(c) When the latitude and time are not accurately known, measure the distance between two stars and compare it with the known apparent distance. The apparent distance is found by the method of § 10, using $a'$, $\delta'$ and $a''$, $\delta''$ as affected by refraction, § 31.

*Example.* The distance between *Aldebaran* and *Arcturus* was measured with the sextant at Ann Arbor, Thursday night, 1891 March 5, as below. It is required to form the

CORRECTIONS TO SEXTANT READINGS     99

equation (177) for this pair of stars. The chronometer correction $\Delta\theta$ was + $15^m\ 7^s$.

| Chronometer | Sextant | Barom. 29.400 inches |
|---|---|---|
| $8^h\ 37^m\ 25^s$ | 130° 14′ 55″ | Att. Therm. 65°.0 F. |
| 45 30 | 14 40 | Ext. Therm. 18°.0 F. |
| 48 30 | 14 55 | |
| 52 20 | 14 35 | Amer. Ephem., pp. 322, 340 |
| 56 20 | 14 55 | *Aldebaran*    *Arcturus* |
| 9 0 30 | 14 45 | α  $4^h\ 29^m\ 39^s\ .33$    $14^h\ 10^m\ 41^s\ .97$ |
| 9 4 0 | 14 40 | δ  16° 17′ 22″ .7    19° 44′ 47″ .8 |
| Means 8 52 5 | 130 14 46.4 | |
| $\Delta\theta$ + 15 7 | | |
| $\theta$ 9 7 12 | | |

With these data we solve (41), (35), (36), (37), (32), (95), (100) and (101) as below.

|  | *Aldebaran* | *Arcturus* |  |  |  |
|---|---|---|---|---|---|
| $\theta$ | $9^h\ 7^m\ 12^s$ | $9^h\ 7^m\ 12^s$ | sin $t$ | $9.97127$ | $9.98667_n$ |
| a | 4 29 39 | 14 10 42 | cos $\phi$ | 9.86915 | 9.86915 |
| $t$ | 4 37 33 | 18 56 30 | cosec $z$ | 0.04642 | 0.03729 |
| $t$ | 69° 23′ 15″ | 284° 7′ 30″ | cosec $q$ | 0.11316 | $0.10689_n$ |
| $\phi$ | 42 16 47 | 42 16 47 | log 1 | 0.00000 | 0.00000 |
| cot $\phi$ | 0.04130 | 0.04130 | True $z$ | 63° 58′ 41″ | 66° 35′ 42″ |
| cos $t$ | 9.54660 | 9.38746 | Mean refr. | 1 56 | 2 11 |
| $L$ | 21° 9′ 54″ | 15° 1′ 24″ | App. $z$ | 63 56 45 | 66° 33 31 |
| δ | 16 17 23 | 19 44 48 | log $\mu$ | 1.75821 | 1.75766 |
| tan $t$ | 0.42467 | $0.59921_n$ | tan $z$ | 0.31078 | 0.36292 |
| sin $L$ | 9.55758 | 9.41366 | $A$ log $BT$ | 9.99583 | 9.99583 |
| sec($\delta+L$) | 0.10027 | 0.08542 | $\lambda$ log $\gamma$ | 0.02746 | 0.02750 |
| tan $q$ | 0.08252 | $0.09829_n$ | log $r$ | 2.09228 | 2.14391 |
| $q$ | 50° 24′ 39″ | 308° 34′ 16″ | sin $q$ | 9.88684 | $9.89311_n$ |
| cot($\delta+L$) | 0.11573 | 0.15849 | sec δ | 0.01780 | 0.02632 |
| cos $q$ | 9.80433 | 9.79482 | log $d$a | 1.99692 | $2.06331_n$ |
| $z$ | 63° 58′ 41″ | 66° 35′ 42″ | cos $q$ | 9.80433 | 9.79482 |
|  |  |  | $d$δ | + 1′ 18″.8 | + 1′ 26″.8 |
|  |  |  | $d$a | + 1 39 .3 | − 1 55 .7 |

Applying these refractions to the above star places we obtain the coördinates which are to be used in solving (53), (54), (55) and (50).

100     PRACTICAL ASTRONOMY

| | | | |
|---|---|---|---|
| $a'$ | 67° 26' 29".2 | $\cot(\delta'' + G)$ | $9.914333_n$ |
| $\delta'$ | 16 18 41 .5 | $\cos B'$ | 9.842600 |
| $a''$ | 212 38 33 .8 | $d$ | 130° 17' 22".4 |
| $\delta''$ | 19 46 14 .6 | | |
| $a'' - a'$ | 145 12 4 .6 | $\sin(a'' - a')$ | 9.756404 |
| $\cot \delta'$ | 0.533068 | $\cos \delta'$ | 9.982158 |
| $\cos(a'' - a')$ | $9.014429_n$ | $\operatorname{cosec} B'$ | 0.143841 |
| $G$ | 280° 36' 52".7 | $\operatorname{cosec} d$ | 0.117597 |
| $\sin G$ | $9.974038_n$ | log 1 | 0.000000 |
| $\tan(a'' - a')$ | $9.841975_n$ | | |
| $\sec(\delta'' + G)$ | 0.107546 | $d'$ | 130° 14' 46".4 |
| $\tan B'$ | 0.013559 | $d - d'$ | + 2 36 .0 |
| $B'$ | 45° 53' 39".3 | $d - d'$ | + 156 .0 |

The angle $\nu$ in (177) is not $\tfrac{1}{2} d'$, but one-half the reading corresponding to the line of the circle with which the vernier line coincides, and it is the eccentricity of this point which enters into $d - d'$. For the reading $d' = 130°\ 14'\ 50''$ the 29th line of the vernier coincides with the circle line 135° 0', and therefore in this case $\tfrac{1}{2} \nu = 33°\ 45'$. We now find

$$\sin \tfrac{1}{2} \nu \cos \tfrac{1}{2} \nu = 0.462, \quad \sin^2 \tfrac{1}{2} \nu = 0.309;$$

and therefore

$$0.462\, x - 0.309\, y + I = 156.0.$$

Similarly, from the meridian double altitude of a star, method (b), and from another pair of stars we find

$$0.259\, x - 0.072\, y + I = 165.0,$$
$$0.117\, x - 0.014\, y + I = 171.0.$$

Solving these three equations we obtain

$$\log x = 1.61380_n, \quad \log y = 0.43265, \quad I = 175.8;$$

whence, from (176),

$$\eta = 176°\ 14', \quad 4\, e'' = 41''.2.$$

While the index correction varies from day to day, and its value should be determined by the methods of § 74 every time the sextant is used, the eccentricity is practically constant. By neglecting the term $I$ in (175) and making $2\nu$ successively 0°, 10°, etc., we obtain the following corrections for eccentricity to be applied to the circle readings.

| Circle | Correction | Circle | Correction | Circle | Correction |
|---|---|---|---|---|---|
| 0° | 0″.0 | 50° | − 8″.8 | 100° | − 16″.2 |
| 10 | − 1 .8 | 60 | − 10 .5 | 110 | − 17 .4 |
| 20 | − 3 .6 | 70 | − 12 .0 | 120 | − 18 .5 |
| 30 | − 5 .4 | 80 | − 13 .5 | 130 | − 19 .4 |
| 40 | − 7 .2 | 90 | − 14 .9 | 140 | − 20 .2 |

In order to determine the eccentricity very accurately, at least ten known angles distributed uniformly from 0° to 140° should be measured and the resulting equations solved by the method of least squares. The observations should be made in one night, so that $I$ may be considered constant; but the observer should determine $I$ several times during the night, to make sure that it does not change.

DETERMINATION OF TIME

**76. Time is determined** from observations on the heavenly bodies by determining the *corrections* to the chronometer or other time-piece at the instants when the observations are made.

**77.** *By equal altitudes of a fixed star.* When a star is from two to four hours east of the meridian and near the prime vertical, observe a series of its double altitudes [§ 71] with the sextant and sidereal chronometer, and let the mean of the chronometer times be $\theta'$. When the star reaches the same altitude west of the meridian observe its double altitudes with the vernier of the sextant set at the same readings as before, in inverted order, and let the mean of the chronometer times be $\theta''$. The chronometer time of the star's meridian passage is $\frac{1}{2}(\theta' + \theta'')$. The sidereal time of the star's meridian passage equals its right ascension $a$. The chronometer correction $\Delta\theta$ at this instant is given by

$$\Delta\theta = a - \tfrac{1}{2}(\theta' + \theta''). \qquad (178)$$

If a mean time chronometer is employed the sidereal time $a$ must be converted into mean time. The required chronometer correction is then given by (178) as before.

*Example.* The following equal altitudes of *Arcturus* were observed with a sextant and sidereal chronometer at Ann Arbor, Saturday night, 1891 April 25. Required the chronometer correction.

| Chronometer Star East | Sextant reading | Chronometer Star West |
|---|---|---|
| $10^h\ 23^m\ 21^s$ | 81° 30′ 0″ | $17^h\ 21^m\ 47^s$ |
| 24 14 | 81 50 0 | 20 52 |
| 25 9 | 82 10 0 | 19 56 |
| 26 4 | 82 30 0 | 19 2 |
| 27 0 | 82 50 0 | 18 7 |
| $\theta'$ 10 25 9.6 | | $\theta''$ 17 19 56.8 |

$$\begin{aligned}
\text{Amer. Ephem., p. 340, } a\quad & 14^h\ 10^m\ 42^s.8 \\
\tfrac{1}{2}(\theta' + \theta'')\quad & 13\ 52\ 33.2 \\
\Delta\theta\quad & +18\ \ \ 9.6
\end{aligned}$$

**78.** *By equal altitudes of the sun.* Observe as described above [§§ 71, 77] the two series of equal double altitudes of the sun before and after noon, and let the chronometer times of the east and west observations be $T'$ and $T''$, a mean time-piece being used. The mean of the two times is not the chronometer time of the sun's meridian passage, since the sun's declination has changed during the interval, and a correction must be applied. To find its value let

$t =$ half the interval between the observations $= \tfrac{1}{2}(T'' - T')$,
$\delta =$ the sun's declination at the observer's apparent noon,
$d\delta =$ the increment of the sun's declination in the interval $t$,
$dt =$ the increment of the sun's hour angle due to the increment of the declination.

DETERMINATION OF TIME   103

Differentiating (15), regarding $\delta$ and $t$ as variables, and dividing by 15 to express $dt$ in seconds of time, we have

$$dt = \left(\frac{\tan \phi}{\sin t} - \frac{\tan \delta}{\tan t}\right)\frac{d\delta}{15}, \quad (179)$$

by which amount the east and west observed times are greater than they would be if the declination were constant and equal to $\delta$. The chronometer time of the sun's meridian passage is therefore $\frac{1}{2}(T' + T'') - dt$. The mean time of the sun's meridian passage is $E$, the equation of time at the observer's apparent noon; and therefore the chronometer correction at the mean of the two times is

$$\Delta T = E - \tfrac{1}{2}(T' + T'') + dt. \quad (180)$$

If a sidereal chronometer is employed the sidereal interval $t$ is converted into the equivalent mean interval [§ 16], $dt$ is computed from (179) as before and subtracted from the mean of the two times, and the result is the chronometer time of the sun's meridian passage. The sidereal time at this instant is equal to the sun's apparent right ascension $a$, and the chronometer correction is given by

$$\Delta\theta = a - \tfrac{1}{2}(\theta' + \theta'') + dt. \quad (181)$$

*Example.* The equal double altitudes of the sun were observed as below at Ann Arbor, Saturday, 1891 April 24–25, a sidereal chronometer being used. Find the chronometer correction.

| Chronometer Sun East | Sextant reading | Chronometer Sun West | Sun's limb |
|---|---|---|---|
| 22ʰ 5ᵐ 4ˢ | 66° 46′ 0″ | 5ʰ 42ᵐ 21ˢ | Upper |
| 5 48 | 67 2 0 | 41 37 | " |
| 6 33 | 67 18 0 | 40 53 | " |
| 7 17 | 67 34 0 | 40 9 | " |
| 8 1 | 67 50 0 | 39 25 | " |
| 9 8.5 | 67 10 0 | 38 17 | Lower |
| 9 53 | 67 26 0 | 37 32 | " |
| 10 38 | 67 42 0 | 36 47.5 | " |
| 11 23 | 67 58 0 | 36 3 | " |
| 22 12 7.5 | 68 14 0 | 5 35 18.5 | " |
| $\theta'$ 22 8 35.3 | | $\theta''$ 5 38 50.3 | |

104   PRACTICAL ASTRONOMY

| | | | |
|---|---|---|---|
| $\frac{1}{2}(\theta'' - \theta') = t$ | $3^h\ 45^m\ 7^s.5$ | tan $\phi$ | 9.9587 |
| Reduction | $-36.9$ | sin $t$ | 9.9192 |
| Mean interval $t$ | $3\ 44\ 30.6$ | (1) | 1.095 |
| $t$ | $3^h.742$ | tan $\delta$ | 9.3725 |
| $t$ | $56°\ 8'$ | tan $t$ | 0.1732 |
| $\phi$ | $42\ 17$ | (2) | 0.158 |
| Amer. Ephem., $\delta$ | $+13\ 16$ | log $[(1)-(2)]$ | 9.9717 |
| $+48''.7 \times 3.742 = d\delta$ | $+182''.2$ | log $d\delta$ | 2.2605 |
| Sun's apparent $a$ | $2^h\ 11^m\ 39^s.3$ | log $\frac{1}{15}$ | 8.8239 |
| $\frac{1}{2}(\theta' + \theta'')$ | $1\ 53\ 42.8$ | log $dt$ | 1.0561 |
| $dt$ | $+11.4$ | | |
| $\Delta\theta$ | $+18\ 7.9$ | | |

**79.** It may be convenient to observe the equal altitudes in the afternoon of one day and the forenoon of the next day. In this case the mean of the two observed times minus the proper value of $dt$ is the chronometer time of the sun's lower culmination. If $t$ is half the mean time interval between the observations it must be replaced by $180 + t = t'$ when substituting in (179); and $E$ in (180) and $a$ in (181) must be increased by $12^h$. The chronometer correction at midnight is then given by (180) and (181).

*Example.* Find the (sidereal) chronometer correction from the following equal altitude observations of the sun.

$\theta'$  $5^h\ 33^m\ 56^s.3$ Friday afternoon, 1891 April 24
$\theta''\ 22\ \ 8\ 35.3$ Saturday forenoon, 1891 April 24

| | | | |
|---|---|---|---|
| $\frac{1}{2}(\theta'' - \theta') = t$ | $8^h\ 17^m\ 19^s.5$ | tan $\phi$ | 9.9587 |
| Reduction | $-1\ 21.5$ | sin $t'$ | $9.9187_n$ |
| Mean interval $t$ | $8\ 15\ 58.0$ | (1) | $-1.096$ |
| $t$ | $8^h.266$ | tan $\delta$ | 9.3668 |
| $t$ | $123°\ 59'$ | tan $t'$ | $0.1713_n$ |
| $t'$ | $303\ 59$ | (2) | $-0.157$ |
| $\phi$ | $42\ 17$ | log $[(1)-(2)]$ | $9.9727_n$ |
| $\delta$ | $+13\ .6$ | log $d\delta$ | 2.6075 |
| $+49''.0 \times 8.266 = d\delta$ | $+405''.0$ | log $\frac{1}{15}$ | 8.8239 |
| $12^h + a$ | $14^h\ 9^m\ 46^s.4$ | log $dt$ | $1.4041_n$ |
| $\frac{1}{2}(\theta' + \theta'')$ | $13\ 51\ 15.8$ | | |
| $dt$ | $-25.4$ | | |
| $\Delta\theta$ | $+18\ 5.2$ | | |

DETERMINATION OF TIME    105

$dt$ is a solar interval, and should be reduced to sidereal, but the correction is small, and for sextant work may be neglected.

80. The method of determining time by equal altitudes possesses the advantages that no corrections are applied for index error, eccentricity, refraction, parallax and semi-diameter; any undetermined errors are eliminated from the result; and the latitude need not be accurately known. However, if the state of the atmosphere and the index correction are different at the two times of observation, the equal sextant readings do not correspond to equal true altitudes, and a correction must be applied. If the index correction is greater and the refraction less for the west observation than for the east, the true double altitude at the west observation is too great by the difference of the index corrections and twice the difference of the refractions, and the time of the observation must be increased by the interval required for the sextant reading to decrease that amount. This interval can be determined from the observations themselves. Thus in the example of § 78, the index correction and refraction for the east observation were

$$I' = +3'\ 8'', \quad r' = 1'\ 24'';$$

and for the west

$$I'' = +3'\ 21'', \quad r'' = 1'\ 22''.$$

The true double altitude at the west observation was too great by

$$(I'' - I') + 2(r' - r'') = +17''.$$

From the observations it is seen that the sextant reading decreased $16' = 960''$ in about $44^s$. If $x$ is the correction to the time of the west observation, we have

$$960 : 17 = 44 : x,$$

from which $x = 0^s.8$. The correction to $\Delta\theta$ is $-\tfrac{1}{2}x = -0^s.4$, and therefore the true value of the chronometer correction is $\Delta\theta = +18^m\ 7^s.5$.

**81. By a single altitude of a star.** A series of double altitudes of a star having been observed in quick succession, let

$R =$ the mean of the sextant readings,
$\theta' =$ the mean of the corresponding chronometer times;

and let

$I =$ the index correction,
$\epsilon =$ the correction for eccentricity,
$h' =$ the apparent altitude of the star,
$z' =$ the apparent zenith distance of the star,
$r =$ the refraction,
$z =$ the true zenith distance of the star.

Then
$$2h' = R + I + \epsilon = 2(90° - z'), \tag{182}$$
and
$$z = z' + r. \tag{183}$$

The latitude $\phi$ and the declination $\delta$ being known, the hour angle $t$ is given by (38) or (39). The sidereal time at the instant of observation is given by $\theta = a + t$, and thence the chronometer correction by

$$\Delta\theta = \theta - \theta'. \tag{184}$$

In case a mean time chronometer is used, the sidereal time $\theta$ must be converted into the mean time $T$ and compared with the chronometer time $T''$.

In determining the time from single altitudes of the stars and the sun, the observations should not be confined to one side of the meridian. It would be well to observe alternately east and west of the meridian, at about equal altitudes. A comparison of the results of such a series often leads to the detection of systematic errors whose presence would not be suspected from observations made wholly in one part of the sky.

*Example.* The observations made on *Arcturus* east of the meridian, recorded in § 77, give

$$\theta' = 10^h\ 25^m\ 9^s.6, \qquad R = 82°\ 10'\ 0''.$$

DETERMINATION OF TIME    107

Find the chronometer correction.

| | | | | | |
|---|---|---|---|---|---|
| R | 82° 10' | 0'' | Barom. | 29.100 inches | |
| I | + 3 | 12 | Att. Therm. | 55° .0 F. | |
| ε | − | 14 | Ext. Therm. | 50 .0 F. | |
| 2h' | 82 12 | 58 | | | |
| h' | 41 6 | 29 | sin ½ [z + (φ − δ)] | 9.76646 | |
| z' | 48 53 | 31 | sin ½ [z − (φ − δ)] | 9.35770 | |
| From (94), r | 1 | 4 | sec ½ [z + (φ + δ)] | 0.24652 | |
| z | 48 54 | 35 | sec ½ [z − (φ + δ)] | 0.00285 | |
| φ | 42 16 | 47 | tan² ½ t | 0.37353 | |
| Ephem., δ | + 19 44 | 52 | tan ½ t | 9.68676ₙ | |
| φ − δ | 22 31 | 55 | ½ t | 154° 4' 26'' | |
| φ + δ | 62 1 | 39 | t | 308 8 52 | |
| z + (φ − δ) | 71 28 | 30 | t | 20ʰ 32ᵐ 35ˢ.5 | |
| z − (φ − δ) | 26 20 | 40 | Ephemeris, α | 14 10 42.7 | |
| z + (φ + δ) | 110 56 | 14 | θ | 10 43 18.2 | |
| z − (φ + δ) | − 13 7 | 4 | θ' | 10 25 9.6 | |
| | | | Δθ | + 18 8.6 | |

**82. By a single altitude of the sun.** If

$p =$ the parallax of the sun,
$S =$ the semidiameter of the sun,

the true zenith distance of the center of the sun is given by

$$z = z' + r - p \pm S; \qquad (185)$$

$S$ being $+$ or $-$ according as the upper or lower limb of the sun was observed. The value of $t$ is given by (38) or (39) as before. $t$ is the true time when the observation was made. The mean time $T$ is given by applying the equation of time $E$. If $T'$ is the chronometer time of observation, the chronometer correction is

$$\Delta T = T - T'. \qquad (186)$$

If a sidereal time-piece is used the mean time $T$ must be converted into the sidereal time and the resulting value compared with the chronometer time.

Since the declination of the sun is changing, it is necessary to know the chronometer correction within 10ˢ; otherwise the value of δ taken from the Ephemeris may be slightly in error, thus giving only an approximate value of the chronometer correction. With this value of the chronometer correction a more accurate value of $\delta$ could be found, which substituted in (38) as before would give practically exact values of $t$ and the chronometer correction.

*Example.* The observations made on the sun east of the meridian, recorded in § 78, give

$$\theta' = 22^h\, 8^m\, 35^s.3, \qquad R = 67°\, 30'\, 0''.$$

The chronometer correction is assumed to be $+18^m\, 3^s$; required its value furnished by the observations.

| | | | | | | |
|---|---|---|---|---|---|---|
| $R$ | 67° | 30' | 0'' | Barom. 29.036 inches | | |
| $I$ | | +3 | 8 | Att. Therm. 50°.0 F. | | |
| $\epsilon$ | | − | 12 | Ext. Therm. 47 .8 F. | | |
| $2h'$ | 67 | 32 | 56 | | | |
| $h'$ | 33 | 46 | 28 | Amer. Ephem., p. 278, $\pi$ | | 8''.8 |
| $z'$ | 56 | 13 | 32 | log $\pi$ | | 0.944 |
| $r$ | | 1 | 24 | sin $z'$ | | 9.920 |
| $p$ | | | 7 | From (64), $p$ | | 7'' |
| $z$ | 56 | 14 | 49 | | | |
| $\phi$ | 42 | 16 | 47 | $\theta$ | $22^h$ | $8^m 35^s$ |
| $\delta$ | +13 | 12 | 53 | Approx. $\Delta\theta$ | +18 | 3 |
| $\phi - \delta$ | 29 | 3 | 54 | Sid. time | 22 26 | 38 |
| $\phi + \delta$ | 55 | 29 | 40 | Mean time April $24^d 20$ | 13 | 29 |
| $z + (\phi - \delta)$ | 85 | 18 | 43 | Longitude | 5 34 | 55 |
| $z - (\phi - \delta)$ | 27 | 10 | 55 | Gr. mean time April 25 | 1 48 | 24 |
| $z + (\phi + \delta)$ | 111 | 44 | 29 | Ephem., $\delta$ | +13°12' | 53'' |
| $z - (\phi + \delta)$ | 0 | 45 | 9 | | | |
| sin $\frac{1}{2}[z + (\phi - \delta)]$ | | | 9.83097 | True time April $24^d 20^h 15^m 38^s.3$ | | |
| sin $\frac{1}{2}[z - (\phi - \delta)]$ | | | 9.37104 | Longitude | 5 34 | 55.1 |
| sec $\frac{1}{2}[z + (\phi + \delta)]$ | | | 0.25099 | Gr. true time April 25 | 1 | 51 |
| sec $\frac{1}{2}[z - (\phi + \delta)]$ | | | 0.00001 | Eq. of time, E | − 2 | 4.9 |
| $\tan^2 \frac{1}{2} t$ | | | 9.45301 | Mean time April 24 | 20 13 | 33.4 |
| $\tan \frac{1}{2} t$ | | | $9.72650_n$ | Sid. time, $\theta$ | 22 26 | 42.4 |
| $\frac{1}{2} t$ | | 151° 57' | 17'' | Chron. time, $\theta'$ | 22 8 | 35.3 |
| $t$ | | 303 54 | 34 | $\Delta\theta$ | +18 | 7.1 |
| True time | | $20^h 15^m 38^s.3$ | | | | |

This value of $\Delta\theta$ differs so little from the assumed value that another approximation to the value of $\delta$ is unnecessary.

Using $+3'\,21''$ as the index correction, the value of $\Delta\theta$ given by the afternoon solar observations, § 78, is $+18^m\,8^s.1$, which agrees well with the above, assuming the chronometer's daily rate to be $+3^s.6$ [§ 64].

**83.** The error in the hour angle — and therefore in the time — produced by a small error in the measured altitude or in the assumed latitude is readily found. Differentiating (15), regarding $z$ and $t$ as variables, and reducing by (17), we obtain

$$dt = \frac{dz}{\sin A \cos \phi}; \qquad (187)$$

that is, an error $dz$ in the measured zenith distance produces an error $dt$ in the time, which is least when $\sin A$ is a maximum. Likewise, differentiating (15) with respect to $\phi$ and $t$ and reducing by (16) and (17), we have

$$dt = -\frac{d\phi}{\tan A \cos \phi}; \qquad (188)$$

that is, an error $d\phi$ in the latitude gives rise to an error $dt$ in the time, which is small when $\tan A$ is large.

For these reasons it appears that to obtain the best determination of time from observed altitudes, those stars should be selected which are as nearly as possible in the prime vertical.

### GEOGRAPHICAL LATITUDE

**84.** *By a meridian altitude of a star or the sun.* Observe the double altitude of the star or sun at the instant when it is on the meridian, and obtain the true zenith distance $z$ as in §§ 81 and 82. The latitude is then found from

$$\phi = \delta \pm z, \qquad (189)$$

the upper sign being used for a star south of the zenith, the lower sign for a star between the zenith and the pole. For a star below the pole we have

$$\phi = 180° - \delta - z. \qquad (190)$$

*Example.* The double altitude of the sun's lower limb was observed at Ann Arbor at true noon, Friday, 1891 Feb. 6, as follows:

Sextant 63° 49′ 15″.  Barom. 28.98 inches,  Ext. Therm. 38° F.

Find the latitude.

In this case $r$ is computed from (96), and $p$ may be taken from the table on page 27.

|   |   |   |   |   |   |   |   |
|---|---|---|---|---|---|---|---|
| $R$ | 63° | 49′ | 15″ | $z'$ | 58° | 3′ | 56″ |
| $I$ | + | 3 | 5 | $r$ |  | 1 | 28 |
| $\epsilon$ | − |  | 12 | $p$ |  |  | 8 |
| $2h'$ |  | 63 | 52 | 8 | $S$ | − | 16 | 15 |
| $h'$ |  | 31 | 56 | 4 | $z$ |  | 57 | 49 | 1 |
| $z'$ |  | 58 | 3 | 56 | $\delta$ | − | 15 | 32 | 11 |
|  |  |  |  |  | $\phi$ |  | 42 | 16 | 50 |

**85.** *By an altitude of a star, the time being known.* Having determined the star's hour angle by (41), the latitude is given by (15), in which $\phi$ is the only unknown quantity. To determine it, assume

$$f \sin F = \cos \delta \cos t, \qquad (191)$$
$$f \cos F = \sin \delta, \qquad (192)$$

and (15) becomes

$$\cos z = f \sin (\phi + F) = \sin \delta \sec F \sin (\phi + F).$$

From these we obtain

$$\tan F = \cot \delta \cos t, \qquad (193)$$
$$\sin (\phi + F) = \cos F \cos z \operatorname{cosec} \delta, \qquad (194)$$

which effect the solution.

The quadrant of $F$ is determined by (191) and (192). $(\phi + F)$, being determined from its sine, may terminate in either of two quadrants, thus giving rise to two values of the latitude. That one is selected which agrees best with the known approximate value of the latitude.

In case the sun is observed, $t$ is the true solar time.

GEOGRAPHICAL LATITUDE   111

*Example.* Find the latitude from the following double altitudes of *Polaris* observed Saturday, 1891 April 25:

| | Chronometer | | Sextant | | Barom. | 29.17 inches |
|---|---|---|---|---|---|---|
| | $14^h 55^m$ | $5^s$ | $82° 18' 40''$ | | Ext. Therm. | $39°.3$ F. |
| | 56 | 35 | 19 20 | | | |
| | 58 | 20 | 19 35 | | Amer. Ephem., p. 305 | |
| | 15 | 1   0 | 20 40 | | $a$ | $1^h 17^m 48^s$ |
| Means | 14 57 | 45 | 82 19 34 | | $\delta$ | $88° 43' 32''$ |
| $\theta'$ | $14^h 57^m$ | $45^s$ | $R$ | $82° 19' 31''$ | $\cot \delta$ | 8.347270 |
| $\Delta\theta$ | $+18$ | 10 | $I$ | $+3\ 12$ | $\cos t$ | $9.939572_n$ |
| $\theta$ | 15 15 | 55 | $\epsilon$ | $-\quad 15$ | $F$ | $358° 53' 28''$ |
| $a$ | 1 17 | 48 | $2h'$ | 82 22 31 | $\cos F$ | 9.999919 |
| $t$ | 13 58 | 7 | $h'$ | 41 11 15 | $\cos z$ | 9.818414 |
| $t$ | $209° 31'$ | $45''$ | $z'$ | 48 48 45 | $\csc \delta$ | 0.000108 |
| | | | $r$ | 1   6 | $\sin(\phi + F)$ | 9.818441 |
| | | | $z$ | 48 49 51 | $\phi + F$ | $41° 10' 20''$ |
| | | | | | $\phi$ | 42 16 52 |

**86.** Differentiating (15) with regard to $z$ and $\phi$ and reducing by (16) we obtain

$$d\phi = \frac{dz}{\cos A};  \qquad (195)$$

that is, an error $dz$ in the measured zenith distance produces the minimum error $d\phi$ in the latitude when the star is on the meridian.

Differentiating (15) with regard to $\phi$ and $t$, and reducing by (16) and (17), we obtain

$$d\phi = - \tan A \cos \phi \, dt;  \qquad (196)$$

that is, an error $dt$ in the estimated time of making the observation gives rise to an error $d\phi$ in the latitude, which will be small when the star is nearly on the meridian, and equal to zero when $A$ is 0° or 180°.

For these reasons it appears that to obtain the best determination of the latitude from observed altitudes, those stars should be selected which are as nearly as possible on the meridian.

**87. By circummeridian altitudes.** The method of § 84 is applicable to only one altitude observed when the star is on the meridian. If a series of altitudes be observed just before and after meridian passage, — called **circummeridian altitudes,** — they can be reduced to the equivalent meridian altitudes and a quite accurate value of the latitude obtained by combining the results. Equation (15) may be written

$$\cos z = \cos(\phi - \delta) - \cos\phi\cos\delta\, 2\sin^2\tfrac{1}{2} t. \tag{197}$$

If we let $z_0$ be the zenith distance of the star when it is on the meridian and put $y = \cos\phi\cos\delta\, 2\sin^2\tfrac{1}{2} t$, (197) becomes

$$\cos z = \cos z_0 - y. \tag{198}$$

Here $z$ is a function of $y$, and we may write

$$z = f(y).$$

Developing this in series by Maclaurin's formula, restoring the value of $y$, and dividing the abstract terms by $\sin 1''$ to express them in seconds of arc, we have

$$z = z_0 + \frac{\cos\phi\cos\delta}{\sin z_0}\cdot\frac{2\sin^2\tfrac{1}{2} t}{\sin 1''} - \left(\frac{\cos\phi\cos\delta}{\sin z_0}\right)^2\cdot\frac{\cot z_0\, 2\sin^4\tfrac{1}{2} t}{\sin 1''} + \cdots, \tag{199}$$

which converges rapidly when $t$ does not exceed $30^m$, and the star is more than $20°$ from the zenith, as it will be in sextant double altitudes. If we let

$$\frac{\cos\phi\cos\delta}{\sin z_0} = A, \qquad A^2\cot z_0 = B, \tag{200}$$

$$\frac{2\sin^2\tfrac{1}{2} t}{\sin 1''} = m, \qquad \frac{2\sin^4\tfrac{1}{2} t}{\sin 1''} = n, \tag{201}$$

and substitute the resulting value of $z_0$ for $z$ in (189), we have

$$\phi = \delta \pm z \mp Am \pm Bn, \tag{202}$$

the lower sign being employed for a star culminating between the zenith and the pole.

When a star is observed near the meridian at lower culmination, it is convenient to reckon the hour angle from the lower transit. $t$ in (15) must be replaced by $180° + t$, and we obtain

$$\cos z = \sin\phi \sin\delta - \cos\phi \cos\delta \cos t = -\cos(\phi+\delta) + \cos\phi \cos\delta \, 2\sin^2 \tfrac{1}{2} t.$$

Developing this in series as before and substituting the resulting value of $z_0$ for $z$ in (190), we have

$$\phi = 180° - \delta - z - Am - Bn. \qquad (203)$$

The entire series of observations is conveniently reduced as a single observation by letting $z$, $m$ and $n$ in (202) and (203) represent the arithmetical means of the values of these quantities for the individual observations.

The values of $m$ and $n$ are tabulated in the Appendix, TABLE III, with the argument $t$.

An approximate value of $\phi$ is required in computing $A$. This may be obtained by the method of § 84, from the observation made nearest the meridian.

If the sun is observed, the declination is taken from the Ephemeris for the instant of each observation in case the observations are reduced separately, and for the mean of the times in case they are reduced collectively.

If a star is observed with a sidereal chronometer, the hour angles $t$ are the intervals between each observed time and the chronometer time of the star's transit.

If a star is observed with a mean time chronometer, the intervals must be reduced from mean to sidereal intervals before entering TABLE III for $m$ and $n$.

If the sun is observed with a mean time chronometer, the intervals should be reduced to apparent solar intervals by correcting for the change in the equation of time during the intervals. This, however, will never exceed $0^s.5$, and may be neglected in sextant observations.

If the sun is observed with a sidereal chronometer, the intervals must be reduced to mean solar intervals and thence to apparent solar.

If the rate of the chronometer is large it must be allowed for.

*Example.* Wednesday, 1891 April 8, at a place in latitude about 42° 17′ and longitude $5^h\ 34^m\ 55^s$ the following double altitudes of the sun were observed with a sextant and sidereal chronometer. Barom. 29.373 inches, Att. Therm. 66° F., Ext. Therm. 42°.5 F. Required the latitude. [Each printed observation is the mean of three consecutive original observations.]

| Limb | Sextant | Chronometer | Sidereal $t$ | Solar $t$ | $m$ | $n$ |
|---|---|---|---|---|---|---|
| Upper | 109° 58′ 33″.7 | 0ʰ 31ᵐ 28ˢ.0 | −19ᵐ 59ˢ.2 | −19ᵐ 55ˢ.9 | 779″.6 | 1″.47 |
| Lower | 109  4 40 .0 | 34 51.0 | −16 36 .2 | −16 33 .5 | 538 .1 | 0 .70 |
| " | 109 14  6 .2 | 38 43 .3 | −12 43 .9 | −12 41 .8 | 316 .4 | 0 .24 |
| Upper | 110 25 16 .2 | 42 42 .3 | − 8 44 .9 | − 8 43 .5 | 149 .5 | 0 .05 |
| " | 110 29 39 .0 | 45 44 .7 | − 5 42 .5 | − 5 41 .6 | 63 .6 | 0 .01 |
| Lower | 109 27 27 .5 | 49 30 .7 | − 1 56 .5 | − 1 56 .2 | 7 .4 | 0 .00 |
| " | 109 27 43 .7 | 53  9 .7 | + 1 42 .5 | + 1 42 .2 | 5 .7 | 0 .00 |
| Upper | 110 29  0 .0 | 0 57 36 .0 | + 6  8 .8 | + 6  7 .8 | 73 .8 | 0 .01 |
| " | 110 21 51 .2 | 1  2 56 .0 | +11 28 .8 | +11 26 .9 | 237 .3 | 0 .16 |
| Lower | 109 11 33 .7 | 5 46 .7 | +14 19 .5 | +14 17 .2 | 400 .6 | 0 .39 |
| " | 109  2 27 .7 | 9 12 .3 | +17 45 .1 | +17 42 .2 | 615 .0 | 0 .92 |
| Upper | 109 56 20 .0 | 1 12 33 .7 | +21  6 .5 | +21  3 .0 | 869 .4 | 1 .83 |
| | 109 45 43 .2 | | | + 0 33 .9 | 339 .7 | 0 .48 |

```
        Apparent time of apparent noon       0ʰ  0ᵐ  0ˢ.0
        Equation of time                       + 1 52.1
        Mean time of apparent noon             0  1 52.1
        Sidereal time of apparent noon         1  8 37.2
        Chronometer correction               +17    10.0
        Chronometer time of apparent noon      0 51 27.2
```

The difference between this and the observed times gives the sidereal intervals $t$ as above.

The mean of the hour angles is $+ 0^m\ 33^s.9$, and therefore the sun's declination is taken for the local mean time $0^h\ 1^m\ 52^s.1 + 0^m\ 33^s.9 = 0^h\ 2^m\ 26^s$, or Greenwich mean time $5^h\ 37^m\ 21^s$.

An equal number of observations on the upper and lower limbs was made, hence there is no correction for semidiameter.

The solution of (202) is made as follows:

GEOGRAPHICAL LONGITUDE 115

| | | | | |
|---|---|---|---|---|
| $\delta$ | $+7° 17' 33''.4$ | Sextant | $109° 45' 43''.2$ | |
| $\phi$ | 42 17 | $I$ | $+2$ 51 .7 | |
| $z_0$ | 34 59 26 .6 | $\epsilon$ | $-$ 18 .0 | |
| $\cos \phi$ | 9.86913 | $2h'$ | 109 48 16 .9 | |
| $\cos \delta$ | 9.99647 | $h'$ | 54 54 8 .4 | |
| $\operatorname{cosec} z_0$ | 0.24151 | $z'$ | 35 5 51 .6 | |
| $\log A$ | 0.10711 | $r'$ | 40 .5 | |
| $\log m$ | 2.53110 | $p$ | 5 .1 | |
| $Am$ | $434''.7$ | $z$ | 35 6 27 .0 | |
| $\log A^2$ | 0.2142 | $\delta$ | $+7$ 17 33 .4 | |
| $\cot z_0$ | 0.1549 | $Am$ | 7 14 .7 | |
| $\log n$ | 9.6812 | $Bn$ | 1 .1 | |
| $\log Bn$ | 0.0503 | $\phi$ | 42 16 46 .8 | |
| $Bn$ | $1''.1$ | | | |

A repetition of the computation with this value of $\phi$ does not change the result.

[The latitude of the place is known to be about $42° 16' 47''.1$.]

GEOGRAPHICAL LONGITUDE

**88.** *By lunar distances.* The moon's distance from a star nearly in the ecliptic is rapidly changing. Its geocentric distances from the sun, *Venus, Mars, Jupiter, Saturn* and nine bright stars near its path are given in the American Ephemeris [pp. XIII–XVIII of each month] at three-hour intervals of Greenwich mean time, from which the distances at any other instants may be found by interpolation. Conversely, if its distance from any of these objects is measured with the sextant and the apparent distance reduced to the corresponding geocentric distance, the Greenwich mean time at the instant of observation may be found. The Greenwich mean time minus the observer's mean time is the observer's longitude. This method of determining the longitude is occasionally of considerable importance to navigators and explorers.

**89.** We shall suppose that the moon's distance from the sun has been observed. The formulæ for a planet will be the same, save that the semidiameter of the planet may usually * be neglected. For a star the parallax and semidiameter are zero.

The sextant reading having been corrected for the index error and eccentricity, the result is the apparent distance between the nearest limbs of the sun and moon. It must be corrected for their semidiameters, refractions and parallaxes.

To compute these corrections, the zenith distances of the two bodies must be known. When there are three observers, as frequently happens at sea, the altitudes of the sun and moon, and the distance between them, should be measured simultaneously. The observer's mean time can be obtained also from these observed altitudes of the sun [§ 82]. When it is not practicable to make these observations at the same time, the observer may measure the altitudes immediately before and after measuring the lunar distance, and obtain the required altitudes at the instant of observation by interpolation. Again, the observer may *assume an approximate value of the longitude* (which he can usually do sufficiently accurately), and take from the Ephemeris the right ascensions and declinations of the sun and moon corresponding to the Greenwich time thus obtained. The hour angles, azimuths and zenith distances are then given by §§ 8 and 5.

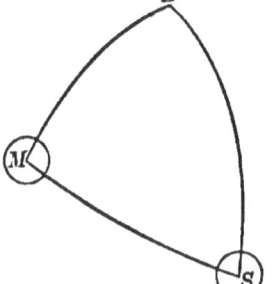

Fig. 18

The parallax of the sun in azimuth is negligible; its parallax in zenith distance is given by (64). The parallax of the moon in azimuth is given by (71) and (72); and in zenith distance, by (80), (78) and (79). (95) gives the refractions, care being taken to use the *apparent* zenith distance. The semidiameters of the sun and

---

* In case the telescope is powerful enough to define the planet's disk, the moon's limb may be made to pass through the center of the disk.

GEOGRAPHICAL LONGITUDE    117

moon are obtained by the methods of §§ 33-35. The solution of (107) requires the values of $q$. In Fig. 18 let $M$ be the moon's center, $S$ the sun's center, and $Z$ the zenith. For the sun, $q = ZSM$, and for the moon, $q = ZMS$. If we let

$Z'$ = the apparent zenith distance of the sun = $ZS$,
$z'$ = the apparent zenith distance of the moon = $ZM$,
$d'$ = the apparent distance between the centers = $SM$,

we can write, for the sun,

$$\tan \tfrac{1}{2} q = \sqrt{\frac{\sin \tfrac{1}{2}(z' - Z' + d') \sin \tfrac{1}{2}(z' + Z' - d')}{\sin \tfrac{1}{2}(Z' + z' + d') \sin \tfrac{1}{2}(Z' - z' + d')}}; \quad (204)$$

and for the moon,

$$\tan \tfrac{1}{2} q = \sqrt{\frac{\sin \tfrac{1}{2}(Z' - z' + d') \sin \tfrac{1}{2}(Z' + z' - d')}{\sin \tfrac{1}{2}(z' + Z' + d') \sin \tfrac{1}{2}(z' - Z' + d')}}. \quad (205)$$

Adding the inclined semidiameters given by (107) to the corrected sextant reading, the sum is the distance $d'$ between the centers as seen from the observer.

The combined effect of the refraction and the parallax in zenith distance is to shift the bodies in their vertical circles without changing the angle $SZM$ at the zenith, which we shall represent by $V$. If we let

$Z$ = the geocentric zenith distance of the sun,
$z$ = the geocentric zenith distance of the moon,
$d''$ = the corresponding distance between the centers,

we can write

$$\cos d'' = \cos z \cos Z + \sin z \sin Z \cos V, \quad (206)$$
$$\cos d' = \cos z' \cos Z' + \sin z' \sin Z' \cos V. \quad (207)$$

Therefore

$$\frac{\cos d'' - \cos z \cos Z}{\sin z \sin Z} = \frac{\cos d' - \cos z' \cos Z'}{\sin z' \sin Z'};$$

or

$$\frac{\cos d'' - \cos(z + Z)}{\sin z \sin Z} = \frac{\cos d' - \cos(z' + Z')}{\sin z' \sin Z'}. \quad (208)$$

If we put $z' + Z' + d' = 2x$, and substitute

$$\cos d' - \cos(z' + Z') = 2\sin x \sin(x - d'),$$
$$\cos d'' = 1 - 2\sin^2 \tfrac{1}{2} d'',$$
$$\cos(z + Z) = 2\cos^2 \tfrac{1}{2}(z + Z) - 1,$$
$$= 1 - 2\sin^2 \tfrac{1}{2}(z + Z),$$

(208) reduces to

$$\sin^2 \tfrac{1}{2} d'' = \sin^2 \tfrac{1}{2}(z + Z) - \frac{\sin z \sin Z}{\sin z' \sin Z'} \sin x \sin(x - d'). \quad (209)$$

Let an auxiliary angle $M$ be defined by

$$\sin^2 M = \frac{\sin z \sin Z}{\sin z' \sin Z'} \cdot \frac{\sin x \sin(x - d')}{\sin^2 \tfrac{1}{2}(z + Z)} \quad (210)$$

Then (209) takes the form

$$\sin \tfrac{1}{2} d'' = \sin \tfrac{1}{2}(z + Z) \cos M. \quad (211)$$

The parallax of the moon in azimuth produces a small change in $V$ and therefore in $d''$. From (206), by differentiation,

$$\Delta d'' = \sin z \sin Z \sin V \operatorname{cosec} d'' \, \Delta V, \quad (212)$$

in which $\Delta V$ is the parallax in azimuth.

The geocentric distance $d$ between the centers is now given by

$$d = d'' + \Delta d''. \quad (213)$$

In connection with the lunar distances, the Ephemeris gives a column "P. L. of Diff." (Proportional Logarithm of the Difference), which is the logarithm of 10800, the number of seconds in $3^h$, minus the logarithm of the change in the lunar distance, expressed in seconds of arc, in the next following three hours. That is, it is the logarithm of the reciprocal of the moon's *average rate* for the three hours, or the rate at the middle period of the three hours [see remarks on interpolation, §15]. In order to interpolate for the Greenwich mean time corresponding to the given value of $d$, we have only to add the P. L. of Diff. for the middle period of the approximate interval to the logarithm of the number of seconds of arc by which $d$ exceeds the next smaller Ephemeris lunar distance. The sum is the logarithm of the number of seconds of time by which the Ephemeris time is to be increased.

If the P. L. of Diff. given in the Ephemeris is used without change, a slight correction for the neglected second difference of the moon's rate can be taken from TABLE I, Appendix, American Ephemeris, and applied as there directed.

If the resulting longitude differs considerably from the assumed longitude, a second approximation should be made by starting with the value of the longitude just obtained. A third approximation will not be necessary.

*Example.* Tuesday, 1891 May 12, the distance between the bright limbs of the sun and moon was observed with a sextant and sidereal chronometer. The mean of ten observations gave

$$\theta' = 8^h\,36^m\,10^s, \quad R = 57°\,28'\,32''.9.$$

Chronometer correction, $+ 19^m\,15^s$; index correction, $+ 2'\,56''.4$; Barom. 29.25 inches, Att. Therm. 62° F., Ext. Therm. 57° F.; latitude, $+ 42°\,16'\,47''$; longitude *assumed*, $+ 5^h\,34^m$. Required a more exact value of the longitude.

|  | | | | | | |
|---|---|---|---|---|---|---|
| $\theta'$ | 8$^h$ 36$^m$ 10$^s$ | | | $R$ | 57° 28' 32''.9 | |
| $\Delta\theta$ | + 19 15 | | | $I$ | + 2 56 .4 | |
| $\theta$ | 8 55 25 | | | $\epsilon$ | − 11 .3 | |
| Mean time | 5 33 42 | | | Distance | 57 31 18 .0 | |
| Longitude | 5 34 | | | | | |
| Gr. mean time | 11 7 42 | | | | | |

Corresponding to this Greenwich mean time, we take from the American Ephemeris, pp. 74, 75, 77, 80 and 278,

| | | Sun | Moon |
|---|---|---|---|
| Right ascension, | $\alpha$ | 3$^h$ 17$^m$ 54$^s$ | 7$^h$ 29$^m$ 31$^s$ |
| Declination, | $\delta$ | + 18° 15' 10'' | + 25° 36' 55'' |
| Semidiameter, | $S$ | 15 51.7 | 15 12.8 |
| Horizontal parallax, | $\pi$ | 8.8 | 55 43.3 |

By §§ 8 and 5 we find for the geocentric coördinates of the sun,

$t = 84° 22' 45''$,   $A = 100° 8' 50''$,   $Z = 73° 46' 5''$;

and for the moon,

$t = 21° 28' 30''$,   $A = 53° 27' 21''$,   $z = 24° 15' 40''$.

Computing the parallaxes we obtain, for the sun,

$A' - A = 0$,   $p = Z' - Z = 8''.4$.

From (58) we find $\phi - \phi' = 687''.3 = 11' 27''.3$; and from (59) $\log \rho = 9.99935$; therefore, for the moon,

$\log m = 6.11807$,   $A' - A = + 21''.8$,

$\gamma = + 6' 49''.2$,   $\log n = 8.20908$,   $z' - z = 23' 6''.2$.

The mean refraction of the sun, TABLE II, is about $3' 13''$, and therefore its apparent zenith distance is very nearly $73° 43' 0''$. The value of the refraction is now found from (95) to be $3' 8''.6$. Similarly, the refraction for the moon is $25''.6$. The apparent zenith distances of the sun and moon are therefore

$Z' = 73° 43' 4''.8$,   $z' = 24° 38' 20''.6$.

The apparent zenith distance of the upper limb of the sun is $73° 43' 4''.8 - 15' 51''.7 = 73° 27' 13''.1$. The corresponding refraction is $3' 5''.5$. The apparent vertical semidiameter is therefore contracted $3''.1$ [§ 35], and its value is $15' 48''.6$.

The moon's apparent semidiameter is found from (106) to be $15' 26''.3$; and by refraction its apparent vertical semidiameter is reduced to $15' 26''.0$.

The approximate distance between the centers of the sun and moon is

$d' = 57° 31' 18'' + 15' 49'' + 15' 26'' = 58° 2' 33''$.

Substituting these values of $d'$, $Z'$ and $z'$ in (204) we obtain for the sun, $q = 20° 58'$; and in (205) for the moon, $q = 124° 34'$. For the sun, $a = 15' 51''.7 = 951''.7$, $b = 15' 48''.6 = 948''.6$; and by (107) the inclined semidiameter is $S'' = 15' 49''.1$. Similarly, for the moon, $S'' = 15' 26''.2$. The apparent distance between the centers of the sun and moon is therefore

$$d' = 57° 31' 18''.0 + 15' 49''.1 + 15' 26''.2 = 58° 2' 33''.3.$$

The solution of (210) gives $M = 49° 48' 39''.0$; and thence, from (211), $d'' = 58° 18' 16''.0$. Substituting $A' - A = \Delta V = + 21''.8$ in (212), we obtain $\Delta d'' = + 7''.4$. The geocentric distance between the sun and moon is therefore, by (213),

$$d = 58° 18' 16''.0 + 7''.4 = 58° 18' 23''.4.$$

From the American Ephemeris, pp. 86 and 87, at

Greenwich mean time  $9^h$, $d = 57° 16' 23''$,    P. L. of Diff. = 0.3169,
Greenwich mean time 12 , $d = 58\ 43\ \ 9$ ,    P. L. of Diff. = 0.3184.

We have to interpolate for the interval of time $T$ after $9^h$, corresponding to a change in $d$ of $58° 18' 23''.4 - 57° 16' 23'' = 3720''.4$. The value of $T$ is approximately $2^h$. The value of P. L. of Diff. at the middle of the $2^h$ is 0.3167.

| | | | |
|---|---|---|---|
| P. L. of Diff. | 0.3167 | Gr. mean time | $11^h\ 8^m 34^s$ |
| log 3720.4 | 3.5706 | Observer's mean time | 5  33  42 |
| log $T$ | 3.8873 | Observer's longitude | 5  34  52 |
| $T$ | $7714^s$ | | |
| $T$ | $2^h\ 8^m\ 34^s$ | | |

The true value of the longitude is known to be $5^h\ 34^m\ 55^s$. The error of $3^s$ corresponds to an error of $2''$ in the measured distance [or in the lunar tables], and is unusually small. The observations are difficult to make, and the measures of the best observers are easily liable to an error of $10''$. It is well, however, to carry the numerous corrections to tenths of a second to prévent the accumulated effect of neglected fractions.

The above solution of this problem is essentially a rigorous one. Navigators are accustomed to employ abridged forms of solution, for which the reductions are much shorter. Likewise many of the functions are tabulated, which still further reduces the labor.

# CHAPTER VII

## THE TRANSIT INSTRUMENT

**90.** The transit instrument consists essentially of a telescope attached perpendicularly to a horizontal axis. The cylindrical extremities of this axis are the **pivots**. The straight line passing through their centers is the **rotation axis**. The supports for the two pivots are called the **V's**. The straight line passing through the optical center of the object glass and the rotation axis and perpendicular to the latter is the **collimation axis**. By revolving the instrument about the rotation axis the collimation axis describes a plane called the **collimation plane**. In the common focus of the object glass and eye-piece is a system of wires called the **recticle**. It consists either of spider threads attached to a frame, or of fine lines ruled on thin glass. An odd number of wires — usually five, seven or eleven — is placed parallel to the collimation plane and perpendicular to the collimation axis, over which the times of transit of a star's image are observed. The **middle wire** of the set is fixed as nearly as possible in the collimation plane. One or two wires are placed perpendicular to these to mark the center of the field of view. A micrometer wire parallel to the first set is arranged to move as nearly as possible in their plane. The axis of the instrument is hollow. A light is placed so that the rays from it enter the axis and fall on a small mirror in the center of the telescope, which reflects them to the eye-piece in such a way that the wires are seen as dark lines in a bright field. The illuminating apparatus in some instruments is arranged so that the observer may change from dark lines in a bright field to bright lines in a dark field, — a necessary arrangement when the object to be observed is very faint. A

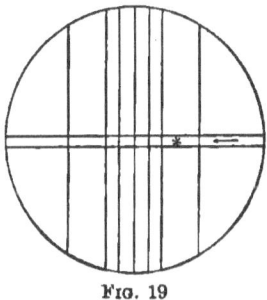

Fig. 19

very common distribution of the wires in the reticle is shown in Fig. 19.

The instrument is so arranged that its rotation axis can be rotated 180°; *i.e.*, *reversed*, about a vertical line. The two positions are defined conveniently by stating the position of the clamp or graduated circle on the axis.
Thus, *clamp W* or *clamp E*, *circle W* or *circle E*, denotes that position of the instrument in which the clamp or circle is west or east of the collimation plane.

An excellent form of transit and zenith telescope *combined* is shown in Fig. 20. The circular base-plate of the instrument is supported on three screws, with which the instrument may be quickly leveled on its supporting pier. The two standards which support the pivots of the instrument are rigidly fixed to a circular plate, which may be rotated freely through 180° on the base-plate when the instrument is used as a zenith telescope [considered in the next chapter]. When the instrument is used as a transit, the two circular plates are clamped together, and remain clamped. To reverse the instrument, the observer turns the reversing crank (shown in the lower right corner of the cut), which raises the two small inner standards until the pivots are entirely free from the V's; the axis, supported on the two standards, is then turned gently through 180°, and carefully lowered by turning the crank in the reverse direction, until the pivots rest in the V's. The illuminating lanterns are in position for the rays to pass through the pivots. The level — in this form called the **striding-level** — is shown resting on the pivots. The micrometer

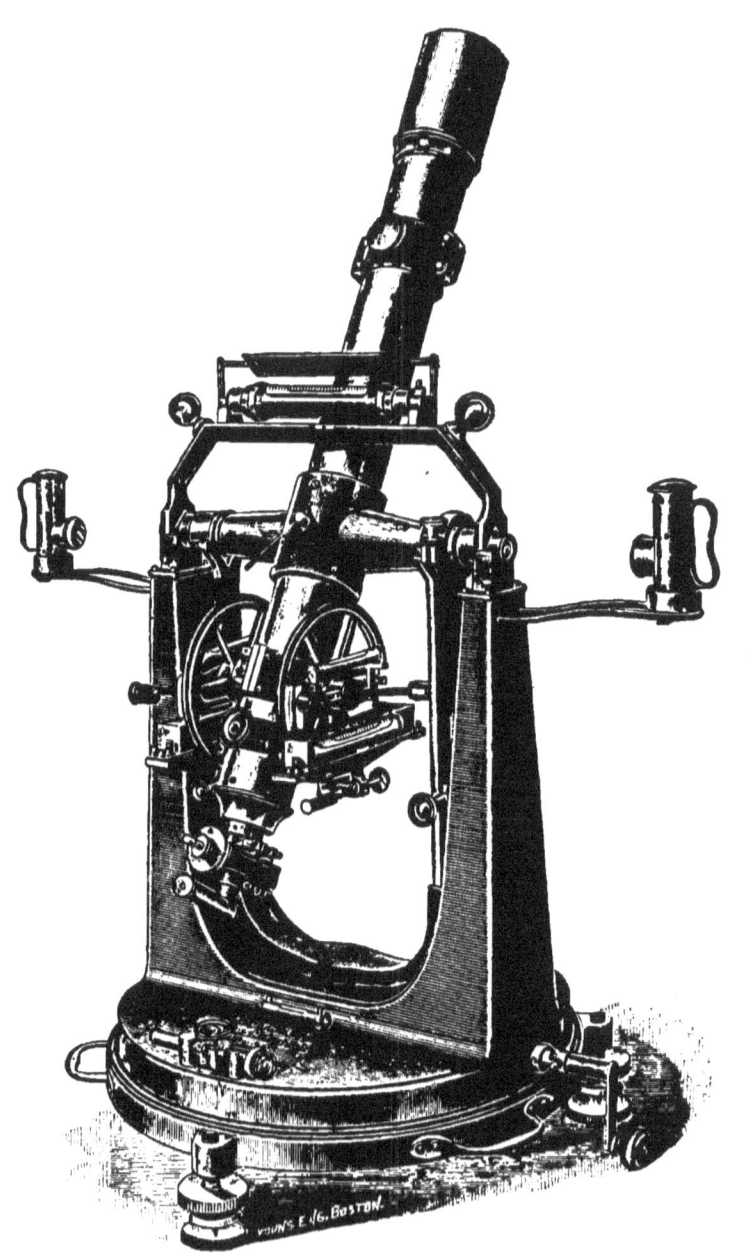

Fig. 20

box can be rotated 90° to make the movable wire perpendicular to the rotation axis for the transit instrument, or parallel to it for the zenith instrument. The cut shows the micrometer in the latter position. A diagonal eyepiece enables the instrument to be used with zenith stars. The small circles attached to the sides of the telescope are used for setting the telescope at the zenith distance or altitude of the star to be observed. The vernier-arms bear both coarse and delicate levels. When one of the verniers is set at the proper reading for the star, the telescope is moved in altitude until the bubble of the coarse level "plays." The star will then pass through the approximate center of the eyepiece. It is made to pass between the two horizontal wires by turning the slow-motion screw.

Another common form of the transit instrument is that in which one end of the axis is made to take the place of the lower half of the telescope. A prism is placed at the intersection of the telescope and axis. This turns the rays of light through 90° to the eyepiece, which is in one end of the axis. This form is sometimes called the **broken** or **prismatic transit.**

An excellent form of the prismatic transit is shown in Fig. 21. In this, the combined telescope and rotation axis is mounted east and west, and the totally reflecting prism is immediately in front of the object glass. It is provided with a reversing apparatus, with a micrometer and delicate zenith level, and can be used also as a zenith telescope. This instrument is very compact, and therefore well adapted for use in exploration or other cases where transportation is difficult.

There are several considerations affecting all forms of the transit instrument; viz.:

The instrument should be reversible without appreciable jarring.

The lamps used for illuminating the reticle, or for

lighting the observing room, must be so placed that they will not heat the instrument appreciably.

The supporting pier must be isolated from the floor of the observing room, and should extend down to a firm rock or soil foundation.

The observing room should be constructed so that it may be thoroughly ventilated before observations are begun.

A sidereal chronometer or clock is a necessary companion of the transit instrument. Refined observations

FIG. 21

should be made by the chronographic method, described in § 68. In a fixed observatory, the clock should not be mounted in the transit room, but in an interior room of more constant temperature, where it can be placed equally well in the electric circuit.

**91.** The transit instrument may be mounted so that its collimation plane is either in the prime vertical, or in the meridian. In the first case it may be used to determine the latitude; but this method is practically superseded by that of the zenith telescope, to be described later. Mounted in the meridian, it is employed in connection with a sidereal clock or chronometer to determine the time, the right ascensions of the stars or other celestial objects, and the longitude of the observer, when great accuracy is required; and we shall treat only this case.

Let us suppose that the axis is mounted due east and west and that the middle wire is exactly in the collimation plane. If the image of a star whose apparent right ascension is $a$ is observed on the wire at the chronometer time $\theta'$, the chronometer correction $\Delta\theta$ is given by (neglecting diurnal aberration)

$$\Delta\theta = a - \theta'. \tag{214}$$

The observer may adjust his instrument as accurately as he pleases, but the adjustments will not remain, owing to changes of temperature, strains, etc. It is customary to put the instrument very nearly in the meridian when it is first set up, and thereafter to vary the adjustments only at long intervals of time. In general, therefore, the star will be observed when it is slightly to one side of the meridian. A determination of the errors of adjustment of his instrument enables the observer to reduce the chronometer time of observation to the chronometer time of meridian passage; whence the chronometer correction is given by (214) as before.

**92.** Theoretically, the rotation axis should be in the prime vertical and in the horizon, and the middle wire should be in the collimation plane.

The **azimuth constant**, $a$, is the angle which the rotation axis makes with the prime vertical. It is + when the west end of the axis is too far south.

The **level constant**, $b$, is the angle which the rotation axis makes with the horizon. It is + when the west end of the axis is too high.

The **collimation constant**, $c$, is the angle which a line through the middle wire and the optical center of the object glass — called the line of sight — makes with the collimation plane. It is + when the middle wire is west (in the eyepiece) of the collimation plane.

It is required to correct the time of observation of a star for the small deviations $a$, $b$ and $c$.

Let $SWNE$ in Fig. 22 represent the celestial sphere projected on the horizon, $Z$ the observer's zenith, $NS$ the meridian, $WE$ the prime vertical, $WQE$ the equator, and $P$

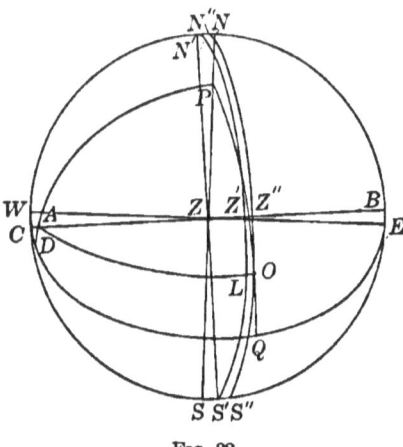

Fig. 22

the pole. Suppose the rotation axis of the instrument lies in the vertical circle $AZB$, and that the axis produced cuts the sphere in $A$ and $B$; that the great circle $N'Z'S'$ lies in the collimation plane; and that $N''Z''S''$, parallel to $N'Z'S'$, is described by the line through the middle wire and the center of the object glass. When the stars are observed on the middle wire they are on the circle $N''Z''S''$, whereas we desire to know the chronometer time when they are on the meridian. Let $O$ be such a star. The time required for the star to pass from $O$ to the meridian is equal to the hour angle of $O$ measured from the meridian *toward the east*. Let $\tau$ represent it.

If we let $90° - m$ denote the hour angle and $n$ the declination of $A$, we have by definition,

THE TRANSIT INSTRUMENT 129

$ZPA = 90° - m,$    $ZA = 90° - b,$
$PZA = 90° + a,$    $PO = 90° - \delta,$
$PA = 90° - n,$     $AO = 90° + c,$
$PZ = 90° - \phi,$  $OPA = 90° - m + \tau.$

From the triangle $ZPA$ we have

$$\sin n = \sin b \sin \phi - \cos b \cos \phi \sin a, \quad (215)$$
$$\sin m \cos n = \sin b \cos \phi + \cos b \sin \phi \sin a; \quad (216)$$

and from $OPA$

$$\sin c = -\sin n \sin \delta + \cos n \cos \delta \sin (\tau - m),$$

or
$$\sin (\tau - m) = \tan n \tan \delta + \sin c \sec n \sec \delta. \quad (217)$$

These equations are true for any position of the instrument, and determine $\tau$ when $a$, $b$ and $c$ are known. But for the instrument nearly in the meridian $a$, $b$, $c$, $m$ and $n$ are small, and the above equations become

$$n = b \sin \phi - a \cos \phi, \quad (218)$$
$$m = b \cos \phi + a \sin \phi, \quad (219)$$
$$\tau = m + n \tan \delta + c \sec \delta. \quad (220)$$

(220) is Bessel's formula for computing the value of $\tau$. Eliminating $m$ and $n$ from the three equations we obtain Mayer's formula

$$\tau = a \cdot \frac{\sin (\phi - \delta)}{\cos \delta} + b \cdot \frac{\cos (\phi - \delta)}{\cos \delta} + c \cdot \frac{1}{\cos \delta}, \quad (221)$$

in which the terms of the second member are the corrections, respectively, for errors of adjustment in azimuth, level and collimation.

For convenience, let us put

$$A = \frac{\sin (\phi - \delta)}{\cos \delta}, \quad B = \frac{\cos (\phi - \delta)}{\cos \delta}, \quad C = \frac{1}{\cos \delta}, \quad (222)$$

and (221) becomes
$$\tau = aA + bB + cC. \quad (223)$$

The effect of the diurnal aberration is to throw the star east of its true position. It is therefore observed too late,

K

and the time of observation must be diminished by the quantity, (116),

$$0''.31 \cos \phi \sec \delta = 0^s.021 \cos \phi \, C. \quad (224)$$

For greater accuracy the star is observed over several wires. An odd number of wires is always used. They are generally placed very nearly equidistant, or very nearly symmetrical with respect to the middle wire. Were either of these arrangements exactly realized the mean of all the times of transit would be the most probable time of transit over the middle wire. This never happens, however, and it is necessary to determine the intervals between the wires.

Let $i$ denote the angular distance between a side wire and the middle wire;* $I$ the interval of time required by a star whose declination is $\delta$ to pass through this distance. From Fig. 13, letting the two positions of the micrometer wire represent the side wire and the middle wire, we have in the triangle $CS'P$,

$$CS' = i, \quad S'P = 90° - \delta, \quad CPS' = I;$$

and we can write

$$\sin I = \sin i \sec \delta = \sin i\, C. \quad (225)$$

If the star is not within 10° of the pole, it is sufficiently accurate to use

$$I = i \sec \delta = i\, C. \quad (226)$$

Suppose there are five threads in the reticle, numbered I, II, III, IV, V, beginning on the side next to the clamp, and that the clamp is west. Let $t_1, t_2, t_3, t_4, t_5$, be the observed times of transit of the star over the wires, and $i_1, i_2, i_4, i_5$, the distances of the four side wires from the middle wire.† The five observed transits give for the time

---

* That is, the angle subtended at the optical center of the object glass by lines drawn to the side wire and to the middle wire. It is also measured by the interval of time required for a star in the equator to pass from the side wire to the middle wire.

† For clamp west, $i_4$ and $i_5$ are negative; for clamp east, $i_1$ and $i_2$ are negative.

of crossing the middle wire either $t_1 + i_1 C$, $t_2 + i_2 C$, $t_3$, $t_4 + i_4 C$, or $t_5 + i_5 C$, which would all be equal if the observations were perfect. Taking their mean, the most probable time of crossing the middle wire is

$$\frac{t_1 + t_2 + t_3 + t_4 + t_5}{5} + \frac{i_1 + i_2 + i_4 + i_5}{5} C.$$

If we let

$$\theta_m = \frac{t_1 + t_2 + t_3 + t_4 + t_5}{5}, \quad (227)$$

and

$$i_m = \frac{i_1 + i_2 + i_4 + i_5}{5}, \quad (228)$$

the most probable time of crossing the middle wire is

$$\theta_m + i_m C. \quad (229)$$

$\theta_m$ is the time of crossing a fictitious wire called the mean wire, and $i_m C$ is the reduction from the mean wire to the middle wire.

The above method holds good also in the case of an *incomplete* transit; that is, one in which the transits over some of the wires have been missed. Thus, suppose that the wires I and IV have been missed. The three remaining transits give for the times of crossing the middle wire $t_2 + i_2 C$, $t_3$, $t_5 + i_5 C$; and their mean is

$$\frac{t_2 + t_3 + t_5}{3} + \frac{i_2 + i_5}{3} C, \quad (230)$$

and similarly in other cases.

In accurate determinations of the time several stars will be observed, and if the chronometer has a sensible rate, the chronometer corrections at the several times of observation will be different. To equalize them, let $\theta_0$ be some chronometer time near the middle of the series of observations, let a star be observed at the time $\theta_m$, and let the rate of the chronometer be $\delta\theta$. During the interval $\theta_m - \theta_0$ the chronomter loses

$$(\theta_m - \theta_0) \delta\theta. \quad (231)$$

If this quantity be computed for all the stars observed and applied to the observed times, the resulting chronometer corrections furnished by the several stars will be the corrections at the instant $\theta_0$, and with perfect observations would all be equal.

Collecting the expressions (223), (224), (229) and (231), we have for the observed time of crossing the meridian when the clamp is west,

$$\theta' = \theta_m + aA + bB + cC - 0^s.021 \cos\phi\, C + i_m C + (\theta_m - \theta_0)\delta\theta; \quad (232)$$

and therefore, by (214),

$$\Delta\theta = a - [\theta_m + aA + bB + (c - 0^s.021 \cos\phi + i_m) C + (\theta_m - \theta_0)\delta\theta]. \quad (233)$$

For clamp east it is easily seen that $c$ and $i_m$ change sign; otherwise the formula remains the same.

The formula has been deduced for a star observed at upper culmination. For a star observed at lower culmination we have only to replace $\delta$ by $180° - \delta$ in the factors $A$, $B$ and $C$, and they become

$$A = \frac{\sin(\phi + \delta)}{\cos\delta}, \quad B = \frac{\cos(\phi + \delta)}{\cos\delta}, \quad C = -\frac{1}{\cos\delta}. \quad (234)$$

The factors $A$, $B$ and $C$ are readily computed with four-place tables. But when an instrument is set up permanently, as in an observatory, their values should be computed for every degree of declination, and tabulated. For polar distances less than 15° it is convenient to have them tabulated for every ten minutes of declination.

## DETERMINATION OF THE WIRE INTERVALS

**93.** (a) If the instrument is provided with a micrometer in right ascension, set the micrometer wire in succession on each of the fixed wires.* The differences of the microm-

---

\* More accurate readings will be obtained, as in many other cases, by setting the micrometer wire on each side of the fixed wire and just in apparent contact with it. The mean of the readings in the two positions is the reading for the coincidence of the two wires.

eter readings on the side wires and the middle wire give the intervals in terms of one revolution of the screw, which will have been obtained by the methods of § 61.

*Example.* Friday, 1891 Feb. 20. The transit instrument of the Detroit Observatory. Four sets of micrometer readings were made when the micrometer wire was in contact with each side of the fixed wires, to find the wire intervals and $i_m$. The numbers in the last line are the means of all the readings on the corresponding wires. The value of one revolution of the screw is, § 61, $R = 45''.042 = 3^s.003$.

| I | II | III | IV | V |
|---|---|---|---|---|
| 29.966 | 27.443 | 24.891 | 22.361 | 19.797 |
| 30.130 | 27.620 | 25.054 | 22.523 | 19.965 |
| 29.969 | 27.445 | 24.890 | 22.357 | 19.799 |
| 30.131 | 27.621 | 25.058 | 22.527 | 19.970 |
| 29.968 | 27.448 | 24.895 | 22.363 | 19.802 |
| 30.135 | 27.618 | 25.060 | 22.528 | 19.969 |
| 29.968 | 27.445 | 24.898 | 22.356 | 19.797 |
| 30.135 | 27.623 | 25.062 | 22.526 | 19.969 |
| 30.050 | 27.533 | 24.976 | 22.443 | 19.883 |

$i_1 = (30.050 - 24.976) R = + 15_s.237,$
$i_2 = (27.533 - 24.976) R = + 7.679,$
$i_4 = (22.443 - 24.976) R = - 7.607,$
$i_5 = (19.883 - 24.976) R = - 15.294.$

$i_m = + 0^s.003$ for clamp west,
$i_m = - 0.003$ for clamp east.

(*b*) Observe the transits of a close circumpolar star over the several wires, and solve (225) for the intervals $i$. It is convenient and sufficiently accurate to use (225) in the form

$$i = \sin I \frac{\cos \delta}{15 \sin 1''}. \tag{235}$$

Solving as in the case of (159), the resulting values of $i$ will be expressed in seconds of time.

*Example.* Monday, 1891 March 16, λ *Ursæ Minoris* was observed at lower culmination with the transit instrument of the Detroit Observatory, clamp east, as below. Required the wire intervals. From the Amer. Ephem., p. 304, $\delta = 88° 57' 47''.3$.

| Wires | Chronom. | $I$ | $I$ | sin $I$ | $i$ |
|---|---|---|---|---|---|
| I | 7$^h$ 2$^m$ 37$^s$ | − 14$^m$ 3$^s$ | − 3° 30' 45'' | 8.787222$_n$ | − 15$^s$.245 |
| II | 9 35 | − 7 5 | − 1 46 15 | 8.489986$_n$ | − 7 .689 |
| III | 16 40 | | | | |
| IV | 23 40 | + 7 0 | + 1 45 0 | 8.484848 | + 7 .599 |
| V | 30 46 | + 14 6 | + 3 31 30 | 8.788762 | + 15 .299 |

A number of stars should be observed in this way, and the mean of all the results adopted as the wire intervals.

#### DETERMINATION OF THE LEVEL CONSTANT

**94.** The level constant $b$ is generally found by means of a spirit level, as explained in § 62. However, the level is applied to the outer surface of the cylindrical pivots and does not give the inclination of the axis, which passes through their centers, unless their radii are equal.

To determine **the inequality of the pivots** and the method of eliminating it, let $A$ and $B$, Fig. 23, be the

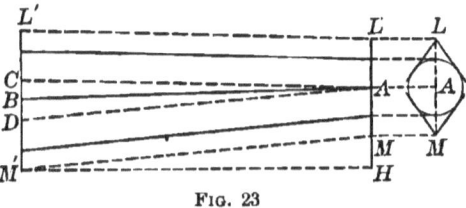

FIG. 23

centers of the west and east pivots, for clamp west; $M$ and $M'$ the vertices of the V's in which the pivots rest; $L$

and $L'$ the vertices of the V's of the level; and $HM'$ a horizontal line. Then $BAC = BAD$ is the inequality of the pivots, which we shall represent by $p$. If we let

$B'$ = the inclination given by the level for clamp west,
$B''$ = the inclination given by the level for clamp east,
$b'$ = the true inclination for clamp west,
$b''$ = the true inclination for clamp east,
$\beta$ = the constant angle $HM'M$,

we can write

$$b' = B' + p = \beta - p, \text{ for clamp west,} \quad (236)$$
$$b'' = B'' - p = \beta + p, \text{ for clamp east;} \quad (237)$$

and therefore

$$p = \frac{B'' - B'}{4}. \quad (238)$$

When $p$ has been determined, the value of the level constant is given by (236) or (237).

The value of $p$ should be determined a large number of times, and the mean of all the individual results adopted as its final value. In making the observations the telescope should be set at different zenith distances, to detect any variations of the pivots from a cylindrical form.

*Example.* 1891 Feb. 17. The following observation was made on the pivots of the Detroit Observatory transit instrument. Required the inequality of the pivots and the inclinations of the rotation axis. The value of one division of the striding level is $d = 1''.878 = 0^s.125$. [See § 63.]

| Clamp | Zenith Distance | Level Direct | | Level Reversed | |
|---|---|---|---|---|---|
| | | $w$ | $e$ | $w'$ | $e'$ |
| W | N. 30° | 7.2 | 9.6 | 15.6 | 1.3 |
| E | S. 30 | 5.2 | 11.7 | 13.5 | 3.4 |

From (166),

$$b = B' = + 2.975\,d = + 0^s.372, \text{ for clamp west};$$
$$b = B'' = + 0.900\,d = + 0^s.112, \text{ for clamp east}.$$

Substituting these values in (238), we find

$$p = - 0^s.065;$$

and therefore, from (236) and (237),

$$b' = + 0^s.372 - 0^s.065 = + 0^s.307,$$
$$b'' = + 0^s.112 + 0^s.065 = + 0^s.177.$$

The mean of twenty-two determinations of $p$ for this instrument gave $p = - 0^s.066 \pm 0^s.001$.

Another method of determining the level constant is given in § 97, (d).

95. We have supposed that the V's in which the pivots rest and the V's of the level are equal, as is usually the case. If they are unequal, let

$2v$ = the angle of the level V,
$2v_1$ = the angle of the V of the pivot bearing;

and it can be shown that the inequality of the pivots is given by

$$p = \frac{B'' - B'}{2} \cdot \frac{\sin v_1}{\sin v + \sin v_1}. \tag{239}$$

Again, the pivots may not be truly cylindrical. If irregularities are surely found to exist, the instrument should be returned to the maker for improvement; or, if that is not practicable, a table of corrections may be constructed for all possible positions of the telescope. It should be emphasized that observations depending upon faulty pivots are very unsatisfactory. The pivots furnished by the best modern instrument makers seldom show appreciable defects.

**96.** The forms and inequality of the pivots may be investigated very satisfactorily by the Harkness spherometer, shown in Fig. 24. The method of applying it to the determination of irregularities in a supposed circular section of a pivot is apparent. To determine the inequality $p$ of two pivots, let

$D$ = the difference of the spherometer readings on the two pivots,
$P$ = the linear pitch of the screw,
$L$ = the distance between the V's of the transit instrument (expressed in the same units as $P$), and
$2v$ = the angle of the spherometer V's.

Then it can be shown that

$$p = \frac{DP}{15\,L \sin 1''(1 + \sin v)}. \quad (240)$$

Fig. 24

## DETERMINATION OF THE COLLIMATION CONSTANT

**97.** (*a*) *By a distant terrestrial object.* Place the telescope in a horizontal position and select some well-defined distant point whose image is seen near the middle wire. With the micrometer measure the distance of the image from the middle wire in the two positions of the instrument. Call this distance $D'$ for clamp west, $D''$ for clamp east; $D'$ and $D''$ being positive or negative according as the middle wire is west or east (in the eyepiece) of the image. The collimation constant is then given by

$$c = \tfrac{1}{2}(D' - D''), \text{ for clamp west.} \quad (241)$$

For clamp east the sign of $c$ is reversed.

*Example.* Saturday, 1891 April 4. The following observations on a distant object nearly in the horizon were made with the transit instrument of the Detroit Observatory. Required the value of $c$. The value of one revolution of the screw is $3^s.003$.

| Clamp | III | Micrometer on image | Micrometer on III |
|---|---|---|---|
| W | W. of image | 34.067 | 24.792 |
|   |   | .065 | .795 |
|   |   | .058 | .793 |
|   |   | Mean 34.063 | Mean 24.793 |
| E | W. of image | 15.750 | 24.794 |
|   |   | .744 | .792 |
|   |   | .752 | .795 |
|   |   | Mean 15.749 | Mean 24.794 |

We have

$$D' = (34.063 - 24.793) R = + 9.270 R,$$

$$D'' = (24.794 - 15.749) R = + 9.045 R;$$

and therefore, from (241),

$$c = -0.112 R = -0^s.336, \text{ for clamp west.}$$

(b) *By a collimator.* This is an ordinary telescope, preferably of the same size as the observing telescope, placed at the side of the observing room, and mounted on an isolated pier in the line of sight of the observing telescope when that is turned into a horizontal position. Spider threads — usually one vertical and one horizontal — are placed exactly in the principal focus of the collimator. When the wires are suitably illuminated by a light shining through the collimator eyepiece, the rays which radiate from them emerge from the collimator object glass in parallel lines, just as if the threads were situated at an infinite distance. When the transit telescope is directed to the collimator, the observer will see in the focus of his instrument the images of the spider threads in the collimator. The vertical image forms a perfect mark from which to determine the collimation constant, by the same process as that described above (a), for a distant terrestrial object. While the threads in the collimator are virtually at an infinite distance, they are in reality only a few feet from the observer, and thus the atmospheric disturbances which affect observations on a distant terrestrial mark are eliminated.

(c) *By a circumpolar star.* Observe the transit of a close circumpolar star over the first two or three wires; then quickly reverse the instrument and observe the transit over as many of the same wires as possible, being sure to determine the level constant both before and after reversing. Reduce the times of transit in the two positions to the equivalent times of crossing the middle wire. Let $\theta_1$ and $\theta_2$ be these times, and let $b'$ and $b''$ be the level constants for clamp west and clamp east. Then by (233), for clamp west,

$$\Delta\theta = a - \theta_1 - aA - b'B - cC + 0^s.021 \cos\phi\, C;{}^*$$

and for clamp east

$$\Delta\theta = a - \theta_2 - aA - b''B + cC + 0^s.021 \cos\phi\, C.$$

Subtracting and solving for $c$ we obtain

$$c = \tfrac{1}{2}(\theta_2 - \theta_1)\cos\delta + \tfrac{1}{2}(b'' - b')\cos(\phi - \delta). \qquad (242)$$

For lower culmination, $\delta$ being replaced by $180° - \delta$,

$$c = -\tfrac{1}{2}(\theta_2 - \theta_1)\cos\delta - \tfrac{1}{2}(b'' - b')\cos(\phi + \delta). \qquad (243)$$

An example is given in § 103.

(d) *By the nadir.* If the telescope be directed vertically downward to a basin of mercury, and a piece of glass be placed diagonally over and close to the eyepiece in such a way that light from a lamp at one side will be reflected into the telescope, the middle wire and its image reflected from the mercury may be seen near together. Measure with the micrometer the distance between the middle wire and its reflected image. Let $M$ be this distance, and consider it positive when the wire is west (in

---

* The correction for rate will be small compared with the probable error of a transit of a slowly-moving northern star, and may be neglected.

the eyepiece) of its image. If the rotation axis is horizontal we have $M = 2c$; but if there is a level constant $b$, the distance is diminished by $2b$, so that $M = 2c - 2b$; or

$$c = \tfrac{1}{2} M + b. \tag{244}$$

With well-constructed instruments, the collimation constant usually remains practically unchanged during a series of observations. The level constant, on the contrary, sometimes varies rapidly. Further, the spirit level is not always trustworthy. Many excellent observers do not use the striding level, but determine the level constant by the method of the nadir, described above. The collimation constant having been determined by one of the many available methods — usually by the aid of two collimations in the case of large instruments — the value of the level constant is given by (244), thus:

$$b = c - \tfrac{1}{2} M. \tag{245}$$

If we wish to determine both the level and collimation constants by the method of the nadir, we measure the distances of the middle wire from its reflected image in the two positions of the instrument; calling this distance + or − according as the middle wire is west or east of its image. Let

$M'$ = the distance for clamp west,
$M''$ = the distance for clamp east,
$b'$ = the level constant for clamp west,
$b''$ = the level constant for clamp east.

We have, for clamp west,

$$c = \tfrac{1}{2} M' + b',$$

and for clamp east,

$$-c = \tfrac{1}{2} M'' + b''.$$

Therefore

$$c = \tfrac{1}{4}(M' - M'') + \tfrac{1}{2}(b' - b''),$$
$$b' + b'' = -\tfrac{1}{2}(M' + M'').$$

DETERMINATION OF THE COLLIMATION CONSTANT 141

From (236) and (237), $b' - b'' = -2p$. Therefore

$c = \frac{1}{4}(M' - M'') - p$, clamp west, (246)
$c = -\frac{1}{4}(M' - M'') + p$, clamp east, (247)
$b' = -\frac{1}{4}(M' + M'') - p$, clamp west, (248)
$b'' = -\frac{1}{4}(M' + M'') + p$, clamp east. (249)

*Example.* 1891 July 24. The following nadir observations were made with the transit instrument of the Lick Observatory. Required the values of $c$, $b'$ and $b''$.

| Clamp | Micrometer on middle wire | | Micrometer on image | |
|---|---|---|---|---|
| W | | 11.025 | | 10.847 |
| | | .125 | | .852 |
| | Mean | 11.075 | Mean | 10.849 |
| E | | 11.020 | | 11.164 |
| | | .135 | | .166 |
| | Mean | 11.077 | Mean | 11.165 |

The middle wire was east of its image in both cases.

For this instrument, $p = -0^s.021$, $R = 2^s.931$. We have

$M' = -(11.075 - 10.849)R = -0^s.662$,
$M'' = -(11.165 - 11.077)R = -0.258$.

Therefore
$c = -0^s.101 + 0^s.021 = -0^s.080$, clamp west,
$b' = +0.230 + 0.021 = +0.251$,
$b'' = +0.230 - 0.021 = +0.209$.

(e) *By two collimators.* When the observing telescope is large, it is inconvenient and very undesirable to determine the collimation constant by any method which involves reversing. This is avoided by using two collimators, one north and the other south of the instrument, the object glasses of the two collimators being turned toward each other and toward the center of the transit instrument. The view of one collimator from the other collimator is obstructed by the intervening transit instrument; but in large instruments apertures are provided on opposite sides of the enlarged central section of the transit telescope, so that when the telescope is directed to the nadir, and the coverings of the apertures removed, the view is unobstructed. The vertical thread in one collimator and the horizontal thread in the other collimator are usually movable by micrometer screws.

Let the vertical micrometer thread in one collimator be brought into exact coincidence with the fixed vertical thread in the other. The lines of sight of the two collimators will then be exactly parallel, and the two vertical threads, viewed by the transit telescope, will represent objects virtually at an infinite distance and having azimuths differing exactly 180°. Measure the distance $D'$ from the middle wire of the transit reticle to the image of the north collimator thread, and the distance $D''$ from the middle wire to the image of the south collimator thread, calling these distances $+$ or $-$ according as the middle wire is west or east (in the eyepiece) of the collimator images. Then we shall have

$$c = \tfrac{1}{2}(D' + D''). \tag{250}$$

## DETERMINATION OF THE AZIMUTH CONSTANT

**98.** The azimuth constant $a$ can be determined only from observations of stars. Let two stars $(a_1, \delta_1)$ and $(a_2, \delta_2)$ be observed. When all the constants except $a$ have been determined, the times of observation of the two stars can be corrected for all errors save the azimuth. Let $\theta_1$ and $\theta_2$ be the times so corrected. Then (233) reduces for the first star, to

$$\Delta\theta = a_1 - \theta_1 - aA_1,$$

and for the second star, to

$$\Delta\theta = a_2 - \theta_2 - aA_2,$$

$A_1$ and $A_2$ being the values of $A$ corresponding to $\delta_1$ and $\delta_2$. Combining these equations, we obtain

$$a = \frac{(a_1 - \theta_1) - (a_2 - \theta_2)}{A_1 - A_2}. \tag{251}$$

It will be seen that to determine $a$ accurately, all the other constants of the instrument must be well determined, since errors in any one or more of them affect the values of $\theta_1$ and $\theta_2$. If the instrument is not mounted in a very stable manner, the right ascensions $a_1$ and $a_2$ should differ as little as possible. The value of $a$ will be determined best when the denominator $A_1 - A_2$ is as large as possible. If both stars are observed at upper culmination, one should be as far south as possible and the other as near the pole as possible, in which case $A_1$ and $A_2$ will be large and opposite in sign. This condition will be fulfilled still better by observing one star $(a_1, \delta_1)$ at lower culmination and the other $(a_2, \delta_2)$ at upper culmination, both as near the pole as possible and differing nearly $12^h$ in right ascension. In this case $a_1$ must be replaced by $12^h + a_1$ and $\delta_1$ by $180° - \delta_1$ in the various formulæ. Stars observed at lower culmination are marked $S. P.$ (*sub polo*).

## MERIDIAN MARK, OR MIRE

**99.** If a transit instrument is to be used for making long series of observations, as at a fixed observatory, it is well to have a permanent meridian mark, or **mire**, to assist in determining the azimuth constant. The mark consists usually of a minute circular hole in a metal plate mounted on a firm pier at a considerable distance to the north or south of the instrument. An isolated pier in the transit room carries a lens whose center is in the line joining the mark and the center of the transit instrument, the focal length of this lens being equal to the distance of the mark from the lens. When the mark is illuminated by a lamp or electric light [controlled by a switch in the transit room] placed behind the metal plate, the rays which fall on the mire lens will be transmitted as parallel rays to the observing telescope, and the observer will see a well-defined image of the mark in the focus of his instrument.

The focal length of the mire lens should be great, in order that the mark may be at a considerable distance, thereby reducing the angular value of any possible motion of the mark. The mire lens for one of the instruments at Pulkowa has a focal length of 556 feet. One at the Lick Observatory has a focal length of 80 feet. Well mounted mires have been found to be almost constant in azimuth for months at a time.

The azimuth of the transit instrument having been determined from observations of a pair of azimuth stars, by the methods of the preceding section, the azimuth of the mire may be determined by measuring the angle between the mire and the middle wire with the micrometer, and combining the result with the known collimation and azimuth constants. The mean of a long series of such determinations may be adopted as the azimuth of the mire; and thereafter a measure of the angle between the mire and middle wire, combined with the known collimation constant, will determine the azimuth constant. Nevertheless, the observation of star pairs for azimuth should be made as usual, and the results thus obtained combined with those obtained from the mire. The relative weights to be assigned to the results from the two methods will be evident after a short experience with them.

If the mire is mounted at a small angle $I$ above or below the horizon of the instrument, the measured angle between the mire and meridian should, in reality, be multiplied by sec $I$, but that is a constant factor, and with most mires need not be taken into account.

### ADJUSTMENTS

**100.** To set up the instrument, it should first be placed by estimation as nearly as possible in the meridian, and the following adjustments made in the order indicated.

1st. To bring the wires in the common focus of the eyepiece and objective, slide the eyepiece in or out until the wires are perfectly well defined. Then direct the telescope to a very distant terrestrial object, or to a star, and move the tube carrying the wires and eyepiece until an image of the object seen on one of the wires will remain on the wire when the position of the eye is changed. *Polaris* is a good star for this purpose, since its image will move very slowly. When the wires are placed satisfactorily in the focus of the objective, the tube carrying them should be clamped firmly, and remain unmolested indefinitely. Different observers will require only to alter the distance of the eyepiece from the wires in order to bring both star and reticle into focus. This adjustment should be made when the atmosphere is steady.

2d. Make the level constant very nearly zero, testing it by the method of § 94.

3d. To make the wires perpendicular to the axis, direct the telescope to a well-defined mark and bisect it with the middle wire. Adjust the reticle so that the object remains on the wire when the telescope is rotated on its axis. The intersection of the two wires of a collimator furnishes an excellent mark for this purpose.

4th. Test the collimation by the methods of § 97, $(a)$, $(b)$, $(d)$ or $(e)$, and move the reticle sidewise until $c$ is made very small.

5th. To set the finding circle, direct the telescope to a bright star near the zenith, whose declination is $\delta$. When the star enters the field of view move the telescope so that the star describes a diameter of the field, and clamp the instrument. If the circle is designed to give the zenith distances, set it at the reading

$$z = \phi - \delta.$$

It will then read correctly for all other stars, neglecting the refraction.

6th. To adjust the instrument in azimuth, direct the telescope to a star near the zenith whose right ascension is $a_1$. Observe its transit over the middle wire, and let the chronometer time of transit be $\theta_1$. The approximate chronometer correction is

$$\Delta\theta_1 = a_1 - \theta_1.$$

Set the telescope for a circumpolar star whose right ascension $a_2$ is a few minutes greater than $a_1$. It culminates approximately at the chronometer time

$$\theta_2 = a_2 - \Delta\theta_1.$$

Rotate the whole instrument horizontally so that the star is on the middle wire at the instant when the chronometer indicates the time $\theta_2$.

7th. Repeat the 2d adjustment.
8th. Repeat the 6th adjustment.
9th. The final adjustment in azimuth should be tested by the method of § 98.

### DETERMINATION OF TIME

**101.** When the chronometer correction is required to be known very accurately, it is customary to observe the transits of ten or twelve stars. The observing list should be made out very carefully, in advance. Half the stars should be observed with clamp west, the other half with clamp east, since any errors in the adopted values of $i_m$, $p$ and $c$ will be practically eliminated by reversing the instrument. To determine $a$ well, a pair of azimuth stars should be observed before reversing, and another pair after reversing. The remaining stars on the list should be those which culminate near the zenith, or between the zenith and equator; since the zenith stars are affected least by an error in the adopted value of $a$, and the time of transit can be estimated most accurately for the rapidly moving equatorial stars. There is no method of eliminating an error in $b$, and it must be very carefully determined. A good program to follow, with small or medium-sized instruments, is

　　Take the level readings
　　Observe half the stars
　　Take the level readings
　　Reverse the instrument
　　Take the level readings
　　Observe half the stars
　　Take the level readings

If there is time between the stars for making further level readings, they should be made. In reversing, the instrument should be handled very carefully to avoid changing the constants.

**102. Example.** Wednesday, 1891 Feb. 25. The following observing list was prepared and the stars observed by

| No. | Object | Mag. | α | | | δ | | Setting | |
|---|---|---|---|---|---|---|---|---|---|
| (1) | Level | | | | | | | | |
| (2) | π Cephei, S. P. | 4.6 | $23^h$ | $4^m$ | $20^s$ | 74° | 47′.9 | N. 62° | 55′ |
| (3) | δ Leonis | 2.3 | 11 | 8 | 20 | 21 | 7 | S. 21 | 10 |
| (4) | ν Ursæ Majoris | 3.3 | | 12 | 37 | 33 | 41 | S. 8 | 36 |
| (5) | σ Leonis | 4.1 | | 15 | 32 | 6 | 38 | S. 35 | 39 |
| (6) | λ Draconis | 3.3 | | 25 | 0 | 69 | 56 | N. 27 | 39 |
| (7) | Level | | | | | | | | |
| | Reverse | | | | | | | | |
| (8) | Level | | | | | | | | |
| (9) | χ Ursæ Majoris | 3.8 | | 40 | 19 | 48 | 23 | N. 6 | 6 |
| (10) | β Leonis | 2.0 | | 43 | 31 | 15 | 11 | S. 27 | 6 |
| (11) | β Virginis | 3.3 | | 45 | 2 | 2 | 23 | S. 39 | 54 |
| (12) | γ Ursæ Majoris | 2.3 | | 48 | 8 | 54 | 18 | N. 12 | 1 |
| (13) | Level | | | | | | | | |
| (14) | ε Corvi | 3.0 | 12 | 4 | 32 | −22 | 1 | S. 64 | 18 |
| (15) | 4 H. Draconis | 4.6 | | 7 | 12 | 78 | 13 .2 | N. 35 | 56 |
| (16) | Level | | | | | | | | |

the eye and ear method with the transit instrument of the Detroit Observatory, to determine the correction to sidereal chronometer Negus no. 721. The stars were selected from the list in the *Berliner Astronomisches Jahrbuch*, pp.

190-327. For convenience in referring to them in the reductions they are numbered, together with the level observations, in the first column. Their magnitudes are given in the third column. The "Setting" is the reading at which the circle is to be set for observing each star. The circle of this instrument reads zero when the telescope points to the zenith and the degrees are numbered in both directions from the zero. The setting is therefore the zenith distance.

The level observations and their reductions are

| (1) | | (7) | | (8) | | (13) | | (16) | |
|---|---|---|---|---|---|---|---|---|---|
| W. | E. | W. | E. | W. | E. | W. | E. | W. | E. |
| 15.1 | 9.1 | 15.1 | 9.3 | 15.3 | 9.0 | 14.3 | 10.3 | 14.6 | 10.1 |
| 12.0 | 12.1 | 13.7 | 10.8 | 8.6 | 15.9 | 11.0 | 13.5 | 11.0 | 13.5 |
| 12.1 | 12.1 | 13.6 | 10.8 | 9.0 | 15.4 | 10.7 | 13.9 | 11.1 | 13.5 |
| 15.2 | 8.9 | 14.8 | 9.5 | 15.1 | 9.4 | 14.0 | 10.5 | 14.4 | 10.1 |
| $B' + 1.525d$ | | $+ 2.100d$ | | $B'' - 0.212d$ | | $+ 0.225d$ | | $+ 0.487d$ | |
| $B' + 0^s.191$ | | $+ 0^s.262$ | | $B'' - 0^s.026$ | | $+ 0^s.028$ | | $+ 0^s.061$ | |
| $p\ \ - 0^s.066$ | | $- 0^s.066$ | | $p\ \ - 0^s.066$ | | $- 0^s.066$ | | $- 0^s.066$ | |
| $b' + 0^s.125$ | | $+ 0^s.196$ | | $b'' + 0^s.040$ | | $+ 0^s.094$ | | $+ 0^s.127$ | |

The times of transit over the five wires are given below.

| Ol. | Object | I | II | III | IV | V | $\theta_m$ | b |
|---|---|---|---|---|---|---|---|---|
| | | s | s | s | s | h m s | h m s | |
| W | (1) | | | | | | | $+ 0^s.125$ |
| " | (2) | 43.9 | 14.6 | 45.3 | 16.0 | 10 48 47.4 | 10 49 45.44 | .136 |
| " | (3) | 26.8 | 35.0 | 43.0 | 51.1 | 53 59.3 | 53 43.04 | .146 |
| " | (4) | 41.8 | 50.9 | 0.0 | 9.2 | 58 18.3 | 58 0.04 | .156 |
| " | (5) | 40.0 | 47.6 | 55.3 | 3.0 | 11 1 10.8 | 11 0 55.34 | .164 |
| " | (6) | 37.1 | 59.3 | 21.5 | 43.9 | 11 6.1 | 10 21.60 | .188 |
| " | (7) | | | | | | | .196 |
| E | (8) | | | | | | | .040 |
| " | (9) | 5.6 | 54.3 | 42.9 | 31.2 | 25 19.6 | 25 42.72 | .057 |
| " | (10) | 10.2 | 2.4 | 54.5 | 46.5 | 28 38.6 | 28 54.44 | .067 |
| " | (11) | 40.7 | 33.2 | 25.6 | 17.9 | 30 — | 30 29.35 | .072 |
| " | (12) | 57.2 | 44.2 | 31.1 | 18.0 | 33 4.8 | 33 31.06 | .080 |
| " | (13) | | | | | | | .094 |
| " | (14) | 12.1 | 3.9 | 55.6 | 47.4 | 49 39.3 | 49 55.66 | .114 |
| " | (15) | 49.2 | 11.5 | 34.2 | 56.4 | 51 19.4 | 52 34.14 | .120 |
| " | (16) | | | | | | | .127 |

DETERMINATION OF TIME 149

For clamp east, and for lower culmination clamp west, the transits occurred in the order V, IV, III, II, I. The mean of the observed times is given in the column $\theta_m$. The values of $b$ are those found by interpolating for the instant of observation, assuming its value to vary uniformly with the time between two consecutive determinations.

Observations for determining the collimation constant $c$ were made by the method ($a$), § 97, on the preceding afternoon and the following forenoon, which gave for clamp west $c = +0^s.112$ and $c = +0^s.108$, respectively. We shall adopt their mean, $c = +0^s.110$.

The hourly rate of the chronometer was $+0^s.15$. Let it be required to determine the chronometer correction at the chronometer time $\theta_0 = 11^h\ 20^m$, which is approximately the mean of the observation times. The correction for rate is $(\theta_m - 11^h\ 20^m) 0^s.15$.

For convenience, let $c - 0^s.021 \cos\phi + i_m = c'$, and we have

$c' = +0^s.110 - 0^s.015 + 0^s.003 = +0^s.098$, for clamp west,
$c' = -0^s.110 - 0^s.015 - 0^s.003 = -0^s.128$, for clamp east.

Star (11) was observed over only the first four wires. In this case $i_m = -\frac{1}{4}(i_1 + i_2 + i_4) = -3^s.827$, and $c' = -0^s.110 - 0^s.015 - 3^s.827 = -3^s.952$.

The values of $A$, $B$ and $C$ are taken from a table computed for the latitude of the Detroit Observatory. They are used here to three decimal places; for ordinary work two are sufficient. To illustrate the application of (222) and (234), we shall compute $A$, $B$ and $C$ for the stars (2) and (3).

| | (2) | | (2) | | (3) | | (3) |
|---|---|---|---|---|---|---|---|
| $\delta$ | 74° 47'.9 | $A$ | +3.395 | $\delta$ | 21° 7' | $A$ | +0.387 |
| $\phi$ | 42 16 .8 | $B$ | −1.736 | $\phi$ | 42 17 | $B$ | +1.000 |
| $\sin(\phi+\delta)$ | 9.0496 | $C$ | −3.813 | $\sin(\phi-\delta)$ | 9.5576 | $C$ | +1.072 |
| $\cos(\phi+\delta)$ | 9.6582$_n$ | | | $\cos(\phi-\delta)$ | 9.9697 | | |
| $\sec\delta$ | 0.5813 | | | $\sec\delta$ | 0.0302 | | |

The apparent right ascensions are taken as accurately as possible from the *Jahrbuch*. We are now prepared to

150  PRACTICAL ASTRONOMY

fill in the columns $A$, $B$, $C$, Rate, $c'C$, $bB$ and $a$; after which we can determine $a$, and thence $aA$ and $\theta'$.

| Star | $A$ | $B$ | $C$ | Rate | $c'C$ | $bB$ | $aA$ | $\theta'$ | $a$ | $\Delta\theta$ | Wt. |
|---|---|---|---|---|---|---|---|---|---|---|---|
| | | | | $s$ | $s$ | $s$ | $s$ | $h\ m\ s$ | $h\ m\ s$ | $m\ s$ | |
| (2) | +3.395 | −1.736 | −3.813 | −.08 | −0.37 | −.24 | −1.86 | 10 49 43.39 | 11 4 20.01 | +14 36.62 | 0 |
| (3) | + .387 | +1.000 | +1.072 | −.07 | + .11 | +.15 | − .15 | 53 43.08 | 5 19.66 | 36.58 | 2 |
| (4) | + .179 | +1.187 | +1.200 | −.06 | + .12 | +.19 | − .07 | 58 0.22 | 12 36.77 | 36.55 | 2 |
| (5) | + .587 | + .818 | +1.007 | −.05 | + .10 | +.18 | − .28 | 11 0 55.29 | 15 31.78 | 36.49 | 2 |
| (6) | −1.353 | +2.582 | +2.915 | −.02 | + .29 | +.49 | + .54 | 10 22.90 | 24 59.52 | 36.62 | 1 |
| (9) | − .160 | +1.497 | +1.506 | +.01 | − .19 | +.09 | + .06 | 25 42.69 | 40 19.26 | 36.57 | 2 |
| (10) | + .472 | + .922 | +1.036 | +.02 | − .13 | +.06 | − .16 | 28 54.23 | 43 30.82 | 36.59 | 2 |
| (11) | + .642 | + .768 | +1.001 | +.03 | −3.95 | +.06 | − .22 | 30 25.27 | 45 1.77 | 36.50 | 2 |
| (12) | − .357 | +1.676 | +1.714 | +.03 | − .22 | +.18 | + .12 | 33 31.12 | 48 7.67 | 36.55 | 1 |
| (14) | + .972 | + .468 | +1.078 | +.07 | − .14 | +.05 | − .34 | 49 55.30 | 12 4 31.81 | 36.51 | 1 |
| (15) | −2.875 | +3.966 | +4.898 | +.08 | − .63 | +.48 | +1.00 | 52 35.07 | 7 11.58 | +14 36.51 | 0 |

Using stars (2) and (6) to determine $a$ we have

$$a_1 = 11^h\ 4^m 20^s.01, \qquad a_2 = 11^h 24^m 59^s.52,$$
$$\theta_1 = 10\ 49\ 44.75, \qquad \theta_2 = 11\ 10\ 22.36,$$
$$a_1 - \theta_1 = +\ 14\ 35.26, \qquad a_2 - \theta_2 = +\ 14\ 37.16,$$
$$A_1 = +\ 3.395, \qquad A_2 = -\ 1.353;$$

and therefore, from (251), $a = -\ 0^s.400$. Similarly, from (14) and (15) we obtain $a = -\ 0^s.348$. Using these as the values of $a$ for clamp west and east respectively, we form the column $aA$. All the corrections have now been computed. Substituting them in (233) for each star, we obtain the values $\Delta\theta$.

Stars (2) and (15) were observed solely to determine $a$, and the values of $\Delta\theta$ furnished by them will be given a weight 0, in the last column. Assigning a weight 2 to the stars which culminate near the zenith and between the zenith and equator, and a weight 1 to those outside these limits, for the reasons given in § 99, we obtain for the weighted mean of the chronometer corrections,

$$\Delta\theta = +\ 14^m\ 36^s.55 \pm 0^s.009,$$

which we shall adopt as the chronometer correction at the time $\theta_0 = 11^h\ 20^m$.

DETERMINATION OF TIME 151

**103.** To illustrate the determination of $c$ by the method of § 97, ($c$), *Polaris* was observed at lower culmination the same night, as below.

| Polaris | | | | Level | | | |
|---|---|---|---|---|---|---|---|
| Clamp W | V | $12^h\,52^m\,6^s$ | | Clamp W | | Clamp E | |
| " " | IV | 12 57 51 | | W | E | W | E |
| Reversed | | | | 13.9 | 10.2 | 9.9 | 14.0 |
| Clamp E | III | 13 3 22 | | 12.0 | 12.0 | 11.4 | 12.5 |
| " " | IV | 13 9 3 | | 12.1 | 12.0 | 11.4 | 12.4 |
| " " | V | 13 14 54 | | 13.8 | 10.2 | 9.8 | 14.0 |

The intervals of time required for *Polaris* to pass from V to III and from IV to III are given by (225) first putting it in the form

$$\sin I = 15 \sin 1'' \sec \delta \cdot i.$$

The value of $\delta$ was $+88°\,43'\,49''$. Substituting $i_4 = 7^s.607$ and $i_5 = 15^s.294$ successively for $i$ in the formula, we find

$$I_4 = 1°\,25'\,50'' = 5^m\,43^s.3, \qquad I_5 = 2°\,52'\,37'' = 11^m\,30^s.5;$$

and therefore the equivalent times of transit over III are

Clamp W

$12^h\,52^m\,6^s + 11^m\,30^s.5 = 13^h\,3^m\,36^s.5$
$12\ \ 57\ \ 51 + 5\ \ 43.3 = 13\ \ 3\ \ 34.3$

Clamp E

$13^h\,3^m\,22^s \qquad\qquad\ = 13^h\,3^m\,22^s.$
$13\ \ 9\ \ 3 - 5^m\,43^s.3 = 13\ \ 3\ \ 19.7$
$13\ \ 14\ \ 54 - 11\ \ 30.5 = 13\ \ 3\ \ 23.5$

Taking the means for clamp west and clamp east, we obtain

$$\theta_1 = 13^h\,3^m\,35^s.4, \qquad \theta_2 = 13^h\,3^m\,21^s.7.$$

The level constants given by the above observations are

$$b' = +0^s.050, \qquad b'' = -0^s.096.$$

Substituting these in (243) we obtain

$$c = +0^s.091, \text{ clamp west.}$$

## REDUCTION BY THE METHOD OF LEAST SQUARES

**104.** In case the chronometer correction is required with all possible accuracy, the series of transit observations should be reduced by the method of least squares. Let us assume that the level constant, the rate and $i_m$ are accurately determined, and that the chronometer correction, the azimuth constant and the collimation constant are to be obtained from the observations. To avoid dealing with large quantities, let $\Delta\theta_0$ be an approximate value of $\Delta\theta$, and $x$ a small correction to $\Delta\theta_0$, such that

$$\Delta\theta_0 + x = \Delta\theta. \qquad (252)$$

Further, let

$$\Delta\theta_0 + \theta_m + bB - 0^s.021 \cos\phi\, C + i_m C + (\theta_m - \theta_0)\,\delta\theta - a = d. \qquad (253)$$

Then (233) takes the form

$$aA \pm cC + x + d = 0, \qquad (254)$$

the lower sign being for clamp east. A value for $\Delta\theta_0$ having been assumed, all the terms in (253) are known for each star. Therefore, $a$, $c$ and $x$ are the only unknown quantities in (254). Each star furnishes an equation of this form, and their solution by the method of least squares gives the most probable values of $a$, $c$ and $x$; and therefore, by (252), the most probable value of $\Delta\theta$.

**105.** The accuracy with which the time of transit of a star over a wire can be estimated depends upon the power of the instrument and the declination of the star. Assistant Schott of the Coast Survey[*] discussed a large number of observations, and found that the probable error of the observed time of transit over one wire is best represented by

$$\epsilon = \sqrt{(0.063)^2 + (0.036)^2 \tan^2\delta}, \text{ for large instruments,}$$

$$\epsilon = \sqrt{(0.080)^2 + (0.063)^2 \tan^2\delta}, \text{ for small instruments.}$$

---

[*] See U. S. Coast and Geodetic Survey Report for 1880.

The values of $\epsilon$ given in the table below are computed from these for the different values of $\delta$. If 1 be the weight of an observation of an equatorial star, $\epsilon_0$ its probable error, and $p$ the weight of an observation of any other star we have, from theory,

$$p = \frac{\epsilon_0^2}{\epsilon^2}.$$

For large instruments, $\epsilon_0 = 0^s.063$, and for small ones, $\epsilon_0 = 0^s.080$. Substituting the values of $\epsilon$ in this equation we find the following values of $p$.

| δ | Large Instruments | | | Small Instruments | | |
|---|---|---|---|---|---|---|
| | $\epsilon$ | $p$ | $\sqrt{p}$ | $\epsilon$ | $p$ | $\sqrt{p}$ |
| 0° | ± 0$^s$.06 | 1.00 | 1.00 | ± 0$^s$.08 | 1.00 | 1.00 |
| 10 | .06 | 1.00 | 1.00 | .08 | .98 | 1.00 |
| 20 | .06 | .98 | 1.00 | .08 | .92 | .96 |
| 30 | .07 | .91 | .95 | .09 | .83 | .91 |
| 40 | .07 | .82 | .90 | .10 | .70 | .83 |
| 50 | .08 | .69 | .83 | .11 | .53 | .73 |
| 55 | .08 | .61 | .78 | .12 | .44 | .66 |
| 60 | .09 | .51 | .71 | .14 | .34 | .59 |
| 65 | .10 | .40 | .63 | .16 | .26 | .51 |
| 70 | .12 | .29 | .54 | .19 | .18 | .42 |
| 75 | .15 | .18 | .43 | .25 | .10 | .32 |
| 80 | .21 | .09 | .30 | .37 | .05 | .22 |
| 85 | .42 | .02 | .15 | .72 | .01 | .11 |
| 86 | .52 | .015 | .122 | .90 | .008 | .088 |
| 87 | .69 | .008 | .091 | 1.21 | .004 | .066 |
| 88 | 1.03 | .004 | .061 | 1.82 | .002 | .044 |
| 89 | 2.06 | .001 | .031 | 3.70 | .000 | .022 |
| 90 | ∞ | .000 | .000 | ∞ | .000 | .000 |

The observation equations (254) should be multiplied through by the square roots of their respective weights before forming the normal equations. (254) becomes

$$\sqrt{p}\,(aA \pm cC + x + d) = 0. \tag{255}$$

In case some of the wires have been missed, the weight is diminished. If we let

$N$ = the whole number of wires,
$n$ = the number of wires observed,
1 = the factor for an observation over the $N$ wires,
$P$ = the factor for an observation over $n$ wires,

then the weight for an incomplete transit is $pP$.
Assistant Schott found that we should use

$$P = \frac{1 + \frac{1.6}{N}}{1 + \frac{1.6}{n}}, \text{ for large instruments,} \qquad (256)$$

$$P = \frac{1 + \frac{2.0}{N}}{1 + \frac{2.0}{n}}, \text{ for small instruments.} \qquad (257)$$

The following table gives the value of $P$ for reticles containing seven and five wires, for the different values of $n$.

| Large Instruments | | | | Small Instruments | | | |
|---|---|---|---|---|---|---|---|
| $n$ | $P$ | $n$ | $P$ | $n$ | $P$ | $n$ | $P$ |
| 7 | 1.00 | 5 | 1.00 | 7 | 1.00 | 5 | 1.00 |
| 6 | .97 | 4 | .94 | 6 | .96 | 4 | .93 |
| 5 | .93 | 3 | .86 | 5 | .92 | 3 | .84 |
| 4 | .88 | 2 | .73 | 4 | .86 | 2 | .70 |
| 3 | .80 | 1 | .51 | 3 | .77 | 1 | .47 |
| 2 | .68 | | | 2 | .64 | | |
| 1 | .47 | | | 1 | .43 | | |

**106.** We shall now apply these methods to the reduction of the transit observations in § 102.

We shall assume $\Delta\theta_0 = +14^m 36^s.5$. The values of $\theta_m$, $bB$, $(\theta_m - \theta_0)\delta\theta$, and $a$ are obtained as before, and we shall use their values tabulated in § 102. To compute the terms $-0^s.021 \cos\phi\, C$ and $i_m C$, let $-0^s.021 \cos\phi + i_m = c''$. Then

$c'' = -0''.015 + 0''.003 = -0''.012$, for clamp west,
$c'' = -0.015 - 0.003 = -0.018$, for clamp east,
$c'' = -0.015 - 3.827 = -3.842$, for star (11).

The products $c''C$ are given in the table below. The value of $d$ is found for each star by (253). The column $\sqrt{p}$ is taken from the table for the large instruments; but for star (11), which is incomplete, the square root of the weight is found from $pP$.

| Star | $c''C$ | $d$ | $\sqrt{p}$ |
|---|---|---|---|
| (2)  | $+ 0''.05$ | $+ 1''.66$ | 0.43 |
| (3)  | $- .01$    | $- .05$    | .99  |
| (4)  | $- .01$    | $- .11$    | .93  |
| (5)  | $- .01$    | $+ .13$    | 1.00 |
| (6)  | $- .03$    | $- .98$    | .54  |
| (9)  | $- .03$    | $+ .03$    | .84  |
| (10) | $- .02$    | $+ .18$    | 1.00 |
| (11) | $- 3.84$   | $+ .33$    | .97  |
| (12) | $- .03$    | $+ .02$    | .79  |
| (14) | $- .02$    | $+ .45$    | .99  |
| (15) | $- .09$    | $- .47$    | .34  |

Substituting the values of $A$, $C$, $d$ and $\sqrt{p}$ in (255), being careful to change the sign of the $c$ term for clamp east, we have the weighted observation equations

$$\begin{aligned}
+ 1.462\,a - 1.640\,c + 0.43\,x + 0.714 &= 0, \\
+ .383 + 1.061 + .99 - .049 &= 0, \\
+ .166 + 1.116 + .93 - .102 &= 0, \\
+ .587 + 1.007 + 1.00 + .130 &= 0, \\
- .731 + 1.574 + .54 - .529 &= 0, \\
- .134 - 1.267 + .84 + .025 &= 0, \\
+ .472 - 1.036 + 1.00 + .180 &= 0, \\
+ .623 - .971 + .97 + .320 &= 0, \\
- .282 - 1.354 + .79 + .016 &= 0, \\
+ .962 - 1.067 + .99 + .445 &= 0, \\
- .977 - 1.696 + .34 - .160 &= 0.
\end{aligned} \quad (256)$$

The normal equations formed from these are

$$+ 5.780\,a - 2.278\,c + 2.714\,x + 2.331 = 0,$$
$$- 2.278 + 18.020 - 2.505 - 2.807 = 0,\quad (259)$$
$$+ 2.714 - 2.505 + 7.689 + 0.918 = 0.$$

Their solution gives

$$a = -0^s.383, \qquad c = +0^s.115, \qquad x = +0^s.053;$$

and therefore

$$\Delta\theta = \Delta\theta_0 + x = +14^m\,36^s.5 + 0^s.053 = +14^m\,36^s.553.$$

The weights of the quantities just determined are

$$p_a = 4.71, \qquad p_c = 16.80, \qquad p_x = 6.29.$$

Substituting the values of $a$, $c$ and $x$ in (258), we obtain the residuals $\sqrt{p}v$,

$-0.012, -.021, +.011, +.074, -.039, -.024, -.067, +.021, +.010, +.007, +.001.$

The sum of the squares of these is $\Sigma pvv = 0.0134$. The probable error $r_1$ of an observation of weight unity is given by

$$r_1 = \pm\,0.674\sqrt{\frac{\Sigma pvv}{m - q}}, \qquad (260)$$

where $m$ is the number of observation equations, and $q$ is the number of unknown quantities. In this case $m = 11$ and $q = 3$. Therefore $r_1 = \pm\,0^s.028$.

The probable errors of the unknowns are given by

$$r_a = \frac{r_1}{\sqrt{p_a}}, \qquad r_c = \frac{r_1}{\sqrt{p_c}}, \qquad r_x = \frac{r_1}{\sqrt{p_x}}. \qquad (261)$$

Therefore

$$r_a = \pm\,0^s.013, \qquad r_c = \pm\,0^s.007, \qquad r_x = \pm\,0^s.011,$$

and

$$a = -0^s.383 \pm 0^s.013,$$
$$c = +0^s.115 \pm 0^s.007,$$
$$\Delta\theta = +14^m\,36^s.553 \pm 0^s.011.$$

## CORRECTION FOR FLEXURE

**107.** In the broken or prismatic transit instrument (§ 90), a correction for flexure due to the bending of the axis must be applied. The effect of the flexure is to change unequally the positions of the eyepiece and objective, which is the same as changing the inclination of the axis. It can therefore be allowed for by changing the measured inclination $b$, using

$b + f$ for clamp west,
$b - f$ for clamp east,

$f$ being the coefficient of flexure, and the eyepiece being on the clamp end of the axis.

It requires special apparatus to determine $f$ directly, so that unless its value for a particular instrument has been well determined, it is best to reduce all the transit observations by the method of least squares, inserting another unknown quantity $f$, thus:

$$\sqrt{p}(aA \pm fB \pm cC + x + d) = 0. \qquad (262)$$

## PERSONAL EQUATION

**108.** It generally occurs that two observers differ appreciably in their estimates of the time of transit of a star over a wire. Some observers acquire the habit of noting a transit too early, while others note it too late. To illustrate, if a star actually transits at $9^s.5$, one observer may note it systematically at $9^s.7$, whereas another may note it systematically at $9^s.2$. An observer's **absolute personal equation** is the quantity which must be applied to his observed time of transit to produce the actual time of transit. The **relative personal equation** of two observers is the quantity which must be applied to the time of transit noted by one observer to produce the time noted by the other.

The personal equation arises from the observers' habits of

observation, and under uniform conditions may be regarded as sensibly constant for *short periods of time*. The relative personal equation of two most skilful observers, Bessel and Struve, was zero in 1814, but in 1821 it had increased to $0^s.8$ and in 1823 to $1^s.0$. An observer's absolute personal equation will depend very considerably upon the circumstances under which he observes. It will in general be different for observations made with a chronograph and for those made by the eye and ear method; for those made with a clock beating seconds and with a chronometer beating half-seconds; for large and for small instruments; for equatorial and for circumpolar stars; for bright and for faint stars; for stars and for the moon's edge; for different positions of the observer's body; for the observer's different degrees of fatigue; and for other variable circumstances.

It is seldom that an observer's absolute personal equation exerts an injurious effect upon results obtained in completed form from his own observations. But when results obtained by two observers are to be compared or combined, it is often essential that their personal equation be eliminated.

The relative personal equation of two observers $A$ and $B$ may be determined by one of many methods.

(a) Let $A$ observe the transit of a star over the first three or four threads of a transit instrument, and $B$ its transit over the remaining threads. For a second star let the observers alternate, $B$ observing the transits over the first threads, and $A$ over the last threads. When twenty-five or more stars have been observed, let the observations of each be reduced to the corresponding times of transit over the middle wire, by equation (226). The difference of times thus obtained for the two observers will be their relative personal equation. The objection to this method is that the observers are liable to be unduly hurried in exchanging positions at the eyepiece.

DETERMINATION OF LONGITUDE 159

(*b*) Let *A* observe a star's transit over all the threads as usual. Let a second star's transit be observed by *B* as usual. In this manner let the observers alternate until each has observed a long and well selected list of stars for determining the clock correction. Let each reduce his observations as usual. The difference of their clock corrections will be their relative personal equation.

(*c*) Various **personal equation machines** have been devised for measuring personal equation. In these an artificial star is made to cross a field of view arranged with a reticle just as in the transit instrument, and the observer notes the times of transit in the usual manner. The actual times of transit are recorded automatically by an electrical device. The difference of the times determined in the two ways is the observer's absolute equation, provided the machine has no personal equation in making its automatic record. At any rate, the difference of the results thus obtained for two observers is their relative personal equation.

The original programs of observation should always be arranged, if possible, with reference to the *direct elimination* of the personal equation. For example, in the case of longitude determinations, the personal equation of the observers is eliminated by their exchanging places when the program of observations is half completed; and similarly in other cases.

DETERMINATION OF GEOGRAPHICAL LONGITUDE

109. The accurate determination of the difference of longitude of two places requires the accurate determination of the time at each place and a method of comparing these times. One of the following methods of comparison is generally employed.

(*a*) *By Transportation of Chronometers.* Let the eastern place be *E*, the western place *W*, and the difference of their longitude, $L$. Determine the correction $\Delta\theta_e$ and the rate $\delta\theta$ of a chronometer at $E$, at the chronometer time $\theta_e$. Carry the chronometer to $W$, and there determine its correction $\Delta\theta_w$ at the chronometer time $\theta_w$. Then

$\theta_w + \Delta\theta_w =$ correct time at $W$ at chronometer time $\theta_w$;
$\theta_w + \Delta\theta_e + \delta\theta(\theta_w - \theta_e) =$ correct time at $E$ at chronometer time $\theta_w$.

Their difference is

$$L = \Delta\theta_e + \delta\theta(\theta_w - \theta_e) - \Delta\theta_w. \qquad (263)$$

The rate of the chronometer during transportation generally differs from its rate when at rest. The change may be eliminated largely by transporting it in both directions between $E$ and $W$. The rate is also a function of the temperature and the lubrication of the pivots. It has been found that the rate $m$ at any temperature $\vartheta$ can be represented by the formula

$$m = m_0 + k(\vartheta - \vartheta_0) - k't, \qquad (264)$$

in which $\vartheta_0$ is the temperature of best compensation, $m_0$ the rate at that temperature with $t = 0$, $t$ the time measured from that instant, $k$ the temperature coefficient and $k'$ the lubrication coefficient. By determining $m_0$, $k$, $\vartheta_0$ and $k'$ for each chronometer, keeping a record of the temperature during transportation, and transporting several chronometers in both directions, the method yields good results. It should never be employed, however, except when the telegraphic method is impracticable.

(b) *By the Electric Telegraph.* To illustrate the simplest application of the method first, let the observers at $E$ and $W$ determine their chronometer corrections. Next, let the observer at $E$ tap the signal key of the telegraph line joining $E$ and $W$ simultaneously with the beats of his chronometer, and let the observer at $W$ note on his chronometer the times of receiving these signals. In the

DETERMINATION OF LONGITUDE  161

same way let the observer at $W$ send return signals to the observer at $E$. Let

$\theta_e$ = correct time at $E$ of sending signal,
$\theta_w$ = correct time at $W$ of receiving signal,
$\theta_w'$ = correct time at $W$ of sending return signal,
$\theta_e'$ = correct time at $E$ of receiving return signal,
$\mu$ = the transmission time.

Then
$$\theta_e + \mu - L = \theta_w,$$
$$\theta_e' - \mu - L = \theta_w'.$$

Therefore
$$L = \tfrac{1}{2}(\theta_e + \theta_e') - \tfrac{1}{2}(\theta_w + \theta_w'), \qquad (265)$$
$$\mu = \tfrac{1}{2}(\theta_w - \theta_w') - \tfrac{1}{2}(\theta_e - \theta_e'). \qquad (266)$$

There are several small errors affecting the value of $L$ obtained by this method of comparison, viz.:

1st. The personal equation of the observers in sending and receiving the signals;

2d. The time required to close the circuit after the finger touches the key, and to move the armature of the receiving magnet through the space in which it plays — called the **armature time**.

3d. The personal equation of the observers in determining the chronometer corrections, and errors in the right ascensions of the stars employed.

These must be eliminated as far as possible in refined determinations. This is best done by a modification of the above method, called the **method of star signals**. *One clock or chronometer, provided with a break-circuit, is placed in the circuit of the telegraph line, and at each station a chronograph and the signal key of a transit instrument are placed in the same circuit.* The same list of stars is observed at both places, thereby eliminating errors in the right ascensions. When the first star crosses the wires of the transit instrument at $E$, the observer makes the records on *both* chronographs by tapping his

M

key. When the same star reaches the meridian of $W$ the observer there makes a similar record on *both* chronographs, and similarly for the other stars. The observers must also make suitable observations for determining the constants of their instruments and the rate of the clock. Let

$\theta_e$ = the clock time when a star is *on the meridian* of $E$, from the chronograph at $E$,

$\theta_e'$ = the same, taken from the chronograph at $W$,

$\theta_w$ = the clock time when the same star is *on the meridian* of $W$, taken from the chronograph at $E$,

$\theta_w'$ = the same, taken from the chronograph at $W$,

$e$ = the absolute personal equation of the observer at $E$,

$w$ = the absolute personal equation of the observer at $W$,

$\delta\theta$ = the correction for rate in the interval $\theta_w - \theta_e$.

Then

$$\theta_w + \delta\theta + w - \mu - L = \theta_e + e,$$

$$\theta_w' + \delta\theta + w + \mu - L = \theta_e' + e.$$

Therefore

$$L = \tfrac{1}{2}(\theta_w + \theta_w') - \tfrac{1}{2}(\theta_e + \theta_e') + \delta\theta + w - e,$$

which we may write

$$L = L_1 + w - e. \qquad (267)$$

If now the observers exchange places and repeat the observations we shall obtain

$$L = L_2 + e - w, \qquad (268)$$

provided their relative personal equation has not changed. Therefore,

$$L = \tfrac{1}{2}(L_1 + L_2). \qquad (269)$$

Great care must be taken in arranging the circuits to insure that the electric constants are the same at both stations. This condition can be secured by means of a rheostat and galvanometer placed in the circuit at each station. If there is any doubt as to the equality of the constants, any difference in the armature times at the two stations may be eliminated by exchanging the electrical apparatus, along with the observers, at the middle of the series.

If the above conditions are realized, the resulting longitude will be free from all errors except the accidental errors of observation.

The method of star signals requires the exclusive use of the connecting telegraph line for several hours on each observing night. If such an arrangement is impossible, the observers must adopt some practicable method. Thus, if the telegraph line can be used only a few minutes each night, a set of adopted signals can be sent back and forth in such a way as to be recorded on both chronographs. The time at the two stations having been accurately determined, from a carefully selected list of stars, the results obtained by this method are nearly as accurate as those obtained by star signals.

The clock or chronometer should never be placed directly in the circuit joining the two stations, as the current would generally be strong enough either to change its rate or to injure its mechanism. It should be placed in a local circuit of its own, with a current just sufficient to work a relay connecting it with the main circuit.

If so desired, a clock or chronometer may be connected with the circuit at each station, so that the beats of both will be recorded on both chronographs.

It is the custom of the Coast Survey to determine longitudes from observations and signals on ten nights, the observers exchanging places at the middle of the series.

(c) *By the Heliotrope*. In mountainous regions the telegraphic method of determining longitudes is usually unavailable. The difference of longitude of two points in sight of each other can be determined from heliographic signals. The necessity for sending signals in both directions and for the observers exchanging stations will be obvious. The equations involved will be similar to those in method (b).

(d) *By Moon Culminations.* The right ascension of the moon is tabulated in the Ephemeris for every hour of Greenwich mean time, whence its value may be computed for any instant of time at a place whose longitude is known. Conversely, if its right ascension is observed at a given place, the Greenwich time corresponding to this right ascension can be taken from the Ephemeris. The Greenwich time minus the time of observation is the longitude of the observer west of Greenwich.

The right ascension is best observed with a transit instrument in the meridian. An observing list, containing two azimuth stars and four or more stars whose declinations are equal to that of the moon as nearly as possible, is arranged so that the moon is near the middle of the list. The transits of the stars and the moon's bright limb are observed in the usual way. From the star transits, the constants of the instrument and the chronometer correction at the instant of observing the moon are obtained, as before. The distance of the moon's bright limb east of the meridian at the time of observation, $\theta_m$, is given very nearly by [the neglected effect of parallax is small when the instrument is nearly in the meridian]

$$\tau = aA + bB + c'C. \qquad (270)$$

The values of $A$, $B$ and $C$ must be computed for the apparent declination; that is, the geocentric declination minus the parallax, given by (61).

$\tau$ is the time required for a star to pass through the angle $\tau$. If $\Delta a$ is the increase of the moon's right ascension in one mean minute (given in the Ephemeris), the mean time required by the moon to describe the angle $\tau$, is $\tau \dfrac{60}{60 - \Delta a}$. The sidereal interval is therefore

$$\tau \frac{60.164}{60.164 - \Delta a}. \quad \text{Let } M = \frac{60.164}{60.164 - \Delta a}.$$

DETERMINATION OF LONGITUDE    165

The values of log $M$ can be taken from the following table:

| $\Delta a$ | log $M$ | $\Delta a$ | log $M$ | $\Delta a$ | log $M$ | $\Delta a$ | log $M$ |
|---|---|---|---|---|---|---|---|
| 1$^s$.65 | 0.0121 | 1$^s$.95 | 0.0143 | 2$^s$.25 | 0.0166 | 2$^s$.55 | 0.0188 |
| 1.70 | .0124 | 2.00 | .0147 | 2.30 | .0169 | 2.60 | .0192 |
| 1.75 | .0128 | 2.05 | .0151 | 2.35 | .0173 | 2.65 | .0196 |
| 1.80 | .0132 | 2.10 | .0154 | 2.40 | .0177 | 2.70 | .0199 |
| 1.85 | .0136 | 2.15 | .0158 | 2.45 | .0181 | 2.75 | .0203 |
| 1.90 | .0139 | 2.20 | .0162 | 2.50 | .0184 | 2.80 | .0207 |

The " sidereal time of semidiameter passing meridian " is tabulated in the American Ephemeris, pp. 385–392. Let $S$ represent it. The right ascension of the moon's center when on the meridian is equal to the observer's sidereal time $\theta$, and is given by

$$a = \theta = \theta_m + \Delta\theta + \tau M \pm S, \qquad (271)$$

the upper or lower sign being used according as the west or east limb is observed.

*Example.* The moon's east limb and seven stars were observed with the transit instrument of the Detroit Observatory, Saturday, 1891 May 23, to determine the longitude.

The star transits gave

$\Delta\theta = + 15^m 34^s.93$ at chronometer time $16^h 13^m$,
$a = - 0^s.360$,
$b = + 0.674$,
$c' = + 0.100$.

The mean of the observed times of transit of the moon's second limb over the five wires was

$\theta_m = 16^h 12^m 45^s.60$.

The moon's geocentric declination $= - 22° 3'$,
Parallax $=\quad\;\; 0\;51$,
The moon's apparent declination $= - 22\;54$.

Therefore, $A = +\,0.985$, $B = +\,0.455$, $C = +\,1.085$; and $\tau = 0^s.04 = \tau M$. From the Ephemeris, p. 388, $S = 1^m\,9^s.90$. Therefore

$a = \theta = 16^h\,12^m\,45^s.60 + 15^m\,34^s.93 + 0^s.04 - 1^m\,9^s.90 = 16^h\,27^m\,10^s.67$.

From the Ephemeris, p. 83, the right ascension at Greenwich mean time $18^h$ was $16^h\,27^m\,20^s.32$. The difference, $9^s.65$, corresponds to a difference in time of about $4^m$. The average increase of right ascension per minute during this interval was $2^s.3058$. The exact value of the interval before $18^h$ is $9.65 \div 2.3058 = 4^m.185 = 4^m\,11^s.10$. The Greenwich mean time corresponding to the observed value of $a$ was therefore $17^h\,55^m\,48^s.90$. The equivalent sidereal time was $22^h\,2^m\,0^s.38$, and the longitude of the observer was

$L = 22^h\,2^m\,0^s.38 - 16^h\,27^m\,10^s.67 = 5^h\,34^m\,49^s.71$.

Longitudes obtained by this method can be regarded only as approximately correct, for two reasons:

1st. An error in the observed right ascension introduces an error $\dfrac{60}{\Delta a}$ times as great in the resulting longitude;

2d. The tables of the moon's motion are imperfect, and the tabulated right ascensions may be slightly in error. This would introduce an error about $\dfrac{60}{\Delta a}$ as great in a resulting longitude. The above example should be reduced anew when the corrections to the moon's right ascensions for 1891 are published.

# CHAPTER VIII

## THE ZENITH TELESCOPE

110. When a sensitive spirit level at right angles to the rotation axis, and a micrometer with wire moving parallel to the axis are added to the transit instrument, it becomes a **zenith telescope**. The level is called a **zenith level**. The transit instrument and the zenith telescope are frequently combined in this way, as shown in Fig. 20.

### DETERMINATION OF GEOGRAPHICAL LATITUDE

111. The zenith telescope is specially adapted to determining the latitude when great accuracy is required. The method employed is known as **Talcott's method**. It consists in measuring the *difference of the zenith distances of two stars*, one of which culminates south of the zenith and the other north of the zenith. The difference of their zenith distances should not exceed half the diameter of the field of view, to avoid observing near the edge of the field. The difference of their right ascensions should not exceed $15^m$ or $20^m$, to avoid any change in the constants of the instrument between the two halves of the observation; nor should the difference be less than $2^m$ or $3^m$, to avoid undue haste. The zenith distances should never exceed $35°$, to avoid uncertainty in the refractions.

To prepare the observing list, an approximate value of the latitude must be known. This can be found from a map, or from a sextant meridian double altitude (§§ 84–87).

Letting the primes refer to the southern star and the seconds to the northern star, we have

$$\delta' = \phi - z', \quad (272)$$
$$\delta'' = \phi + z''. \quad (273)$$

Therefore

$$\delta' + \delta'' = 2\phi + (z'' - z'), \quad (274)$$

which is the condition that the two stars of the pair must fulfill. Thus, in latitude 42° 17′, and with an instrument whose field of view is 40′ in diameter, we must have two stars such that $\delta' + \delta''$ is greater than 84° 14′ and less than 84° 54′. A pair is given below which meets these requirements. The "Setting" is the mean of the zenith distances. The assumed latitude is 42° 17′.

| Star | Mag. | Apparent α | δ | z | Setting |
|---|---|---|---|---|---|
| κ Ursæ Majoris | 3.3 | $8^h 56^m 12^s$ | + 47° 35′ | N. 5° 18′ | N. 5° 9′ |
| 38 Lyncis | 4.1 | 9 12 5 | 37 16 | S. 5 1 | S. 5 9 |

Care must be taken, in forming the observing list, to employ only those stars whose declinations are well determined.

To observe the first star, the circle to which the zenith level is usually attached is made to read the "Setting," the telescope is rotated until the bubble moves to the middle of the tube, and the micrometer wire is moved to the part of the eyepiece where it is known the star will pass. Thus, in the pair above, it is known that the first star will cross 9′ [ = 5° 18′ − 5° 9′] above the center. When the first star culminates, or within a few seconds of culmination, bisect the star by the micrometer wire, and read the zenith level and the micrometer. Reverse the instrument without jarring it, bring the bubble to the center of the level again, and observe the second star in the same way as the first. It is sometimes preferable not to clamp the

DETERMINATION OF LATITUDE 169

instrument during the observations. *Care must be taken not to change the position of the level with respect to the line of sight during the progress of an observation; the angle between the two must be preserved.*

Let $m_0$ be the micrometer reading on any point of the field assumed as the micrometer zero; $z_0$ the apparent zenith distance corresponding to $m_0$ when the level bubble is at the center of the tube; $m'$, $m''$ the micrometer readings on the two stars, the readings being supposed to increase with the zenith distance; $R$ the value of a revolution of the micrometer screw; $b'$, $b''$ the level constants for the two stars, plus when the north end is high; $r'$, $r''$ the refractions for the two stars. Then the true zenith distance of the southern star is given by

$$z' = z_0 + (m' - m_0) R + b' + r';$$

and of the northern star

$$z'' = z_0 + (m'' - m_0) R - b'' + r''.$$

Substituting these in (274) and solving for $\phi$, we obtain

$$\phi = \tfrac{1}{2}(\delta' + \delta'') + \tfrac{1}{2}(m' - m'')R + \tfrac{1}{2}(b' + b'') + \tfrac{1}{2}(r' - r''). \quad (275)$$

If the micrometer readings decrease for increasing zenith distances the sign of the second term is minus.

In case the zero of the level scale is at the center of the tube,

$$\tfrac{1}{2}(b' + b'') = \tfrac{1}{4}[(n' + n'') - (s' + s')] d, \quad (276)$$

in which $n'$, $n''$, $s'$, $s''$ are the level readings for the two stars, and $d$ is the value of a division of the level.

In case the zero of the level scale is at one end of the tube,

$$\tfrac{1}{2}(b' + b'') = \tfrac{1}{4}[\pm (n' + s') \mp (n'' + s'')] d, \quad (277)$$

the upper sign being used when $n'$ is greater than $s'$, the lower when $n'$ is less than $s'$.

The refraction correction is small, and can be computed differentially by the formula

$$\tfrac{1}{2}(r' - r'') = \tfrac{1}{2}\frac{dr}{dz}(z' - z''), \qquad (278)$$

in which $(z' - z'')$ is expressed in minutes of arc, and $\dfrac{dr}{dz}$ is the rate of change of refraction in seconds of arc per minute of change in zenith distance. Differentiating (97),

$$r = 58'' \tan z,$$

we obtain $\qquad \dfrac{dr}{dz} = 58'' \sec^2 z \sin 1', \qquad (279)$

the factor, sin 1', being introduced to make the two members homogeneous. Therefore, from (278),

$$\tfrac{1}{2}(r' - r'') = 29'' \sec^2 z \sin 1' (z' - z''). \qquad (280)$$

The values of $\tfrac{1}{2}(r' - r'')$ can be taken from the following table, for the mean zenith distance $z$ of the stars. Since the micrometer term of the formula (275) gives the approximate value of $\tfrac{1}{2}(z' - z'')$, this is used as the argument of the table, rather than $z' - z''$. The sign of this correction is the same as that of the micrometer correction.

Values of $\tfrac{1}{2}(r' - r'')$

| $\dfrac{z' - z''}{2}$ | $z = 0°$ | $z = 10°$ | $z = 20°$ | $z = 25°$ | $z = 30°$ | $z = 35°$ |
|---|---|---|---|---|---|---|
| 0' | ".00 | ".00 | ".00 | ".00 | ".00 | ".00 |
| 1 | .02 | .02 | .02 | .02 | .02 | .02 |
| 2 | .03 | .03 | .04 | .04 | .04 | .05 |
| 3 | .05 | .05 | .06 | .06 | .07 | .08 |
| 4 | .07 | .07 | .08 | .08 | .09 | .10 |
| 5 | .08 | .09 | .10 | .10 | .11 | .13 |
| 6 | .10 | .10 | .11 | .12 | .13 | .15 |
| 7 | .12 | .12 | .13 | .14 | .15 | .18 |
| 8 | .13 | .14 | .15 | .16 | .18 | .21 |
| 9 | .15 | .16 | .17 | .18 | .20 | .23 |
| 10 | .17 | .18 | .19 | .21 | .23 | .26 |
| 11 | .18 | .19 | .21 | .23 | .25 | .28 |
| 12 | .20 | .21 | .23 | .25 | .27 | .31 |

In connection with the subject of refraction, a word should be said in regard to securing good observing conditions. The observing room should be thrown open for thorough ventilation an hour or more before the observations begin. The observing room should assume, as nearly as possible, the temperature of the outside air. The line of sight should not pass within the field of influence of a neighboring chimney, or other disturbing factor. Refined observations should not be attempted when the star images are very unsteady.

If for any reason the star cannot be observed at the instant of culmination, the bisection may be made when the star is at some distance from the center of the field, the time of observation being noted. The polar distance of every star observed in this way will be too small. A slight correction, called **the reduction to the meridian**, must be applied. Let $x$ represent it; and let $t$ be the distance of the star from the meridian when it was observed, in seconds of time.

In the right triangle formed by the meridian, the star's declination circle, and the micrometer wire projected on the sphere, we have the side $90° - \delta$ and the angle $t$ at the pole, to find the side $90° - (\delta \pm x)$. We can write

$$\cot(\delta \pm x) = \cos t \cot \delta.$$

Expanding and solving for $\tan x$,

$$\tan x = \pm \frac{(1 - \cos t)\sin \delta \cos \delta}{\sin^2 \delta + \cos t \cos^2 \delta}.$$

We can put the denominator equal to unity without sensible error, since $t$ is always small. Therefore

$$\tan x = \pm 2 \sin^2 \tfrac{1}{2} t \sin \delta \cos \delta = \pm \tfrac{1}{2}\sin 2\delta \cdot 2\sin^2 \tfrac{1}{2} t;$$

or,

$$x = \pm \sin 2\delta \frac{\sin^2 \tfrac{1}{2} t}{\sin 1''}; \tag{281}$$

the lower sign being used for stars observed near lower culmination.

The correction to the observed latitude will always be $\frac{1}{2}x$. If both stars of the pair are observed off the meridian, there will be two such terms to apply.

The values of $x$ are tabulated below with the arguments $\delta$ and $t$.

VALUES OF $x$

| $t \diagdown \delta$ | 5ˢ | 10ˢ | 15ˢ | 20ˢ | 25ˢ | 30ˢ | 35ˢ | 40ˢ | 45 | 50ˢ | 55ˢ | 60ˢ | $t \diagdown \delta$ |
|---|---|---|---|---|---|---|---|---|---|---|---|---|---|
| 0° | ″.00 | ″.00 | ″.00 | ″.00 | ″.00 | ″.00 | ″.00 | ″.00 | ″.00 | ″.00 | ″.00 | ″.00 | 90° |
| 5 | .00 | .00 | .01 | .02 | .03 | .04 | .06 | .08 | .10 | .12 | .14 | .17 | 85 |
| 10 | .00 | .01 | .02 | .04 | .06 | .08 | .11 | .15 | .19 | .23 | .28 | .34 | 80 |
| 15 | .00 | .01 | .03 | .05 | .09 | .12 | .17 | .22 | .28 | .34 | .41 | .49 | 75 |
| 20 | .00 | .02 | .04 | .07 | .11 | .16 | .22 | .28 | .36 | .44 | .53 | .63 | 70 |
| 25 | .01 | .02 | .05 | .08 | .13 | .19 | .26 | .34 | .42 | .52 | .63 | .75 | 65 |
| 30 | .01 | .02 | .05 | .09 | .15 | .21 | .29 | .38 | .48 | .59 | .71 | .85 | 60 |
| 35 | .01 | .03 | .06 | .10 | .16 | .23 | .31 | .41 | .52 | .64 | .77 | .92 | 55 |
| 40 | .01 | .03 | .06 | .11 | .17 | .24 | .33 | .43 | .54 | .67 | .81 | .97 | 50 |
| 45 | .01 | .03 | .06 | .11 | .17 | .25 | .33 | .44 | .55 | .68 | .82 | .98 | 45 |

112. The adjustments for the transit instrument, § 100, apply equally well, for the most part, to the zenith telescope. Special forms of the instrument, however, will call for special methods, which the intelligent observer will easily devise.

The micrometer wire must be made perpendicular to the meridian. If this adjustment is perfect, an equatorial star will travel on the wire throughout its entire length.

113. *Example.* The following observations were made with the zenith telescope of the Detroit Observatory, Monday, 1891 March 16.

| Star | Chronometer | Micrometer | Level | |
|---|---|---|---|---|
| | | | $n$ | $s$ |
| κ *Ursæ Majoris* | 8ʰ 40ᵐ 36ˢ | 13.647 | 8.9 | 35.7 |
| 38 *Lyncis* | 8 56 10 | 37.359 | 39.6 | 12.4 |

Required the latitude.

DETERMINATION OF LATITUDE    173

The chronometer correction was $+15^m\ 47^s$. The value of one revolution of the micrometer screw is $R = 45''.042$. The value of one division of the level is $d = 2''.74$. The mean places of the stars are given in the *Jahrbuch*, p. 180. Their apparent places are found by the methods of § 55 to be

$$a'' = 8^h\ 56^m\ 12^s, \qquad \delta'' = +47°\ 35'\ 21''.33,$$
$$a' = 9\ \ 12\ \ \ 5, \qquad \delta' = +37\ \ 15\ \ 53\ .05.$$

Therefore
$$\tfrac{1}{2}(\delta' + \delta'') = 42°\ 25'\ 37''.19.$$

The micrometer readings decreased with increasing zenith distances. Therefore

$$-\tfrac{1}{2}(m' - m'')R = -11.856\,R = -8'\ 54''.02.$$

The zero of the level was at one end; therefore, by (277),

$$\tfrac{1}{2}(b' + b'') = \tfrac{1}{4}(52.0 - 44.6)d = +5''.07.$$

The half difference of the zenith distances is $8'\ 54''$, and the mean zenith distance is $z = 5°\ 9'$. Therefore, from the table for differential refraction,

$$\tfrac{1}{2}(r' - r'') = -0''.16.$$

The first star was observed at an hour angle $t = +11^s$; therefore, from the table, the value of $\tfrac{1}{2}x$ for the northern star is

$$\tfrac{1}{2}x'' = +0''.02.$$

The second star was observed at the hour angle $t = -8^s$; therefore the value of $\tfrac{1}{2}x$ for the southern star is

$$\tfrac{1}{2}x' = +0''.01.$$

Combining the terms of (275) and the reductions to the meridian, we obtain

$$\phi = 42°\ 16'\ 48''.11.$$

[The known value of the latitude is about $42°\ 16'\ 47''.3$.]

**114.** In very accurate determinations of the latitude, a number of pairs of stars should be observed several times in this way, and the results combined by the method of least squares. If we let $\phi_0$, $R_0$ and $d_0$ be very nearly the true values of $\phi$, $R$ and $d$, and let $\Delta\phi$, $\Delta R$ and $\Delta d$ be slight corrections to $\phi_0$, $R_0$ and $d_0$, each observation furnishes an equation of the form

$$\phi_0 + \Delta\phi = \tfrac{1}{2}(\delta' + \delta'') + \tfrac{1}{2}(m' - m'')(R_0 + \Delta R)$$
$$+ \tfrac{1}{4}[\pm(n' + s') \mp (n'' + s'')](d_0 + \Delta d) + \tfrac{1}{2}(r' - r'') + \tfrac{1}{2}x' + \tfrac{1}{2}x''.$$

Let

$$k = \phi_0 - \tfrac{1}{2}(\delta' + \delta'') - \tfrac{1}{2}(m' - m'')R_0$$
$$- \tfrac{1}{4}[\pm(n' + s') \mp (n'' + s'')]d_0 - \tfrac{1}{2}(r' - r'') - \tfrac{1}{2}x' - \tfrac{1}{2}x''. \quad (282)$$

Then

$$\Delta\phi - \tfrac{1}{2}(m' - m'')\Delta R - \tfrac{1}{4}[\pm(n' + s') \mp (n'' + s'')]\Delta d + k = 0, \quad (283)$$

is an observation equation for determining $\Delta\phi$, $\Delta R$ and $\Delta d$. Thus, in the example above, if we assume $\phi_0 = 42°\ 16'\ 47''.0$, $R_0 = 45''.040$, $d_0 = 2''.70$, we find $k = -1''.06$, and (283) becomes

$$\Delta\phi + 11.856\,\Delta R - 1.85\,\Delta d - 1.06 = 0. \quad (284)$$

Forming the corresponding equations for the other pairs observed and solving by the method of least squares, the most probable values of $\Delta\phi$, $\Delta R$ and $\Delta d$, and therefore of $\phi$, $R$ and $d$, are obtained.

However, it has recently been shown that the latitude of a place varies appreciably, sometimes in the course of a few weeks; and latitude observations, to be combined by any direct method, must be made inside of a few days and the result be taken as the latitude at the mean of the observation times.

# CHAPTER IX

## THE MERIDIAN CIRCLE

**115. The Meridian Circle** consists essentially of a transit instrument with a graduated circle attached at right angles to, and concentric with, the rotation axis. The graduated circle rotates in common with the telescope, and is read by reading microscopes firmly attached to one of the supporting piers of the instrument. The best instruments are provided with two graduated circles. One of these circles usually remains fixed on the axis for an indefinite time; whereas the other is movable, and many observers are accustomed to rotate it through any desired angle [and clamp it firmly to the axis] from time to time.

An excellent form of the meridian circle is illustrated, with many details omitted, in Fig. 25. Two massive supporting piers extend down to solid earth or rock foundation; and, as in the case of all telescope piers, are completely isolated from the floor and building. The circular drum on each pier carries the four long slender reading microscopes for reading the graduated circle, on which they are focused. The pivots rest in V's attached to the inner head plates of the drums. The counterbalance levers are shown on the tops of the drums. A lifting arm descends from the inner end of each lever and rests, with roller bearings, on the under side of the axis. The chain descending from the outer end of the lever, through a hole in the pier, carries a counterweight. Nearly the whole weight of the instrument is supported in this manner, leaving only a small residual weight to be borne by the V's.

FIG. 25

The level is in position on the instrument, suspended from the pivots. It is visible immediately under the circles. The eyepiece is supplied with the usual number of vertical threads, and with vertical and horizontal micrometer wires. A basin of mercury, not visible in the cut, is mounted below the level of the floor, immediately under the center of the instrument, on a pier isolated from the floor. The small telescope showing just below the objective of the lower left reading microscope is the "setting telescope," used for setting the instrument at any desired circle reading. The auxiliary apparatus shown is, the observing chair in the foreground, the adjustable mercury basin for reflection observations, and the reversing carriage in the background. The field of view and the wires are illuminated by light from a lamp at the side of the room, shining through the hollow axis, as in the case of the transit instrument. A system of small mirrors receives light from the same source and reflects it where it is needed for reading the circles and setting the telescope. The graduated circles are about two feet in diameter, and the graduations are two minutes of arc apart. The micrometer head of each microscope is divided into 60 parts, each division corresponding to $1''$. The divisions may be subdivided, by estimation, into 10 parts, each part being $0''.1$.

The meridian circle is used principally to determine the accurate positions of the heavenly bodies on the celestial sphere; *i.e.*, their right ascensions and declinations. It is further adapted to determining the time and the geographical longitude by the methods of Chapter VII, and to determining the geographical latitude.

**116.** *The determination of right ascensions.* The principles involved in this problem have been treated in Chapter VII. The stars whose right ascensions are to be determined (which we shall call **undetermined stars**), are placed in an observing program which includes a considerable number of stars whose positions are accurately known (which we shall call **standard stars**), and which are suitable for determining the azimuth of the instrument, and the time. The transits of all the stars, both undetermined and standard, are observed, the constants of the instrument are determined, and all the observations are reduced in the usual manner. The clock correction is determined from the standard stars, as usual. The right ascensions of the undetermined stars are found by means of equation (233), which may be written

$$a = \theta_m + \Delta\theta + aA + bB + c'C + (\theta_m - \theta_0)\delta\theta. \qquad (285)$$

In forming the observing program, the undetermined stars should be preceded, accompanied and followed by standard stars. Likewise, the declinations of the standard stars should be nearly the same as, or at least should include, the declinations of the undetermined stars, thereby eliminating largely the uncertainties or progressive changes in the instrumental constants.

*Example.* In the example of § 102, let it be assumed that the star (3) $\delta$ *Leonis* and star (10) $\beta$ *Leonis* are undetermined stars, and that the remaining nine stars of the list are standard stars. Required the right ascensions of stars (3) and (10).

The value of the chronometer correction from the nine standard stars is

$$\Delta\theta = + 14^m 36^s.54.$$

The right ascension computations may now be tabulated, as below.

|  | Star (3) $\delta$ *Leonis* | Star (10) $\beta$ *Leonis* |
|---|---|---|
| $\theta_m$ | $10^h 53^m 43^s.04$ | $11^h 28^m 54^s.44$ |
| $\Delta\theta$ | + 14 36.54 | + 14 36.54 |
| $aA$ | − 0.15 | − 0.16 |
| $bB$ | + 0.15 | + 0.06 |
| $c'C$ | + 0.11 | − 0.13 |
| $(\theta_m - \theta_0)\delta\theta$ | − 0.07 | + 0.02 |
| $a$ | 11  8  19.62 | 11  43  30.77 |

**117.** Declinations and the latitude are determined from observations which involve readings of the graduated circle. The method of reading the circle by reading microscopes, and correcting for error of runs, is given in § 58. In modern instruments, provided with eyepiece micrometers moving in declination, the error of runs may and should be practically eliminated by a suitable method of observing. To illustrate, if it is the observer's custom to read each microscope on only one graduation, the telescope should be directed, by means of the setting telescope, so that the graduation to be set on will always fall in the same position with reference to the zero of the microscope for all the observations of a series. Again, if it is the observer's custom to read each microscope on two adjacent graduations, these should always fall in the same positions with reference to the micrometer zero, the two graduations being, preferably, on opposite sides of the zero. Knowing the approximate position of the star to be observed, its "setting"—usually zenith distance—can be computed to the nearest *even* minute, and the instrument set for that reading. As the star crosses the middle thread in the eyepiece, its distance from the zero position of the micrometer wire is measured with the declination micrometer. The microscope readings on the graduations are then secured, and later the declination micrometer is read.

In reading the circle it is customary to take the degrees and *even* minutes from the circle as seen in one of the microscopes, and the seconds and fractions from the mean of the four microscope readings. The reading thus obtained must be corrected for runs, for flexure, for the distance of the declination micrometer wire from its zero position, and possibly for errors of the graduations.

**118.** The zero reading of the micrometer may be obtained from nadir observations [§ 97, (*d*)]. Let the observing telescope be directed vertically downward to the mercurial basin. Obtain the micrometer readings when the wire is on each side of its reflected image, at minute and equal distances. The mean of the two readings is the reading for coincidence of the wire and its image, and is the **zero reading** of the micrometer. Let the microscopes be read for this position of the instrument, and corrected for runs and graduation error. The result is the **nadir reading** of the circle. The nadir reading plus 180° is the **zenith reading**.

Many of the older forms of meridian circles are not provided with declination micrometers, but have two horizontal fixed wires marking the center of the field, as shown in Fig. 19. In this case, when a star is crossing the middle transit thread, the entire instrument is moved by a slow motion screw until the star travels midway between the two horizontal wires. The microscope readings may thus have any value up to 2', and the correction for runs must be carefully determined. Again, some instruments have a single horizontal fixed wire.

**119.** *Determination of the value of one revolution of the micrometer screw.* The method of § 61, (*a*), is applicable, but a better method is the following: Direct the observing telescope to one of the collimators (described in § 97), so that the image of the horizontal wire of the collimator falls about half a radius above the center of the field. Determine the micrometer reading when the micrometer wire is coincident with the image of the collimator wire, and read the circle. Rotate the instrument so that the collimator image moves to the opposite side of the field of view, and again determine the corresponding micrometer and circle readings. The difference of the circle readings divided by the difference of the micrometer readings is the value of a revolution of the screw. If a movable circle is available, several different arcs may be used for this purpose, thereby eliminating very largely the effect of graduation errors.

THE MERIDIAN CIRCLE 181

**120.** *Eccentricity of the circle.* As explained in § 59, the effect of eccentric mounting of the circle is eliminated by the use of two or more equidistant verniers or reading microscopes.

**121.** *Flexure.* When the instrument is rotated from one position to another, the form of the observing telescope (and sometimes also the form of the graduated circle), is appreciably changed under the action of gravity. The bending of the telescope tube will do no harm provided the objective and the eyepiece are displaced the same amount, but a difference in their displacements changes the direction of the line of sight with reference to the circle graduations. This effect is called the **flexure**.

In most modern instruments the flexure is very small, since the observing telescope is symmetrical with reference to the rotation axis, and the mechanism at the eye end is made of the same weight as the objective and its cell. They are further designed so that the objective and eyepiece mechanism may be interchanged on the telescope tube. If we combine two observations of the same body, one made before interchanging the objective and ocular, and the other after interchanging them (the interchange involving at the same time a rotation of the telescope tube through 180°), the result will be free from flexure, theoretically at least.

The two collimators furnish a simple method of measuring the **horizontal flexure**, *i.e.*, the flexure when the telescope is in a horizontal position. Let the horizontal threads of the collimators be brought into coincidence, as explained in § 97, (*e*). The two threads then represent two infinitely distant lines whose angular distance apart, measured through the zenith, is exactly 180°. Measure this distance in the usual manner. If there is no flexure, the difference of the circle readings should be exactly 180°. If any excess or deficiency exists, that excess or deficiency is

182    PRACTICAL ASTRONOMY

twice the horizontal flexure, plus the accidental and unavoidable errors of the observation.

*Example.* Repsold Meridian Circle, Lick Observatory, Saturday, 1898 June 11, the following observations were made by R. H. Tucker for determining the horizontal flexure. Circle east.

| | |
|---|---|
| Circle Reading on South Collimator . . . . | 224° 56′ 49″.07 |
| North Collimator set on South Collimator | |
| Circle Reading on North Collimator . . . . | 44 56 48 .64 |
| North Collimator set on South Collimator | |
| Circle Reading on North Collimator . . . . | 44 56 48 .58 |
| Circle Reading on South Collimator . . . . | 224 56 48 .75 |
| Mean Circle Reading, North . . . . . . . . | 44 56 48 .61 |
| Mean Circle Reading, South . . . . . . . . | 224 56 48 .91 |
| Difference, North–South . . . . . . . . . | 179 59 59 .70 |

The deficiency is $0''.30$ and the horizontal flexure, $f$, is $0''.15$. The sign of the flexure correction to the circle readings is readily found. As the telescope was turned from the south collimator through the zenith to the north collimator, the readings increased from 224° through 360° to 44°, and the measure of the angle is $0''.30$ too small. The correction to the circle reading for a star south of the zenith is minus, and for a star north of the zenith is plus. If the instrument were reversed, circle west, the signs of the corrections would be reversed.

The mean value of $f$, resulting from 23 determinations by the same observer, extending through two years, is $0''.065$, but the value $0''.1$ has been adopted, provisionally.

Since the gravitational moment of any given mass in the telescope, with reference to the rotation axis, varies with the sine of the zenith distance of the line of sight, the general expression for the flexure is assumed to be

$$\text{Flexure} = f \sin z, \qquad (286)$$

though it is not probable that the flexures in all instruments can be represented by this law.

The order of observation followed in the above example illustrates a general principle which should be taken into account, whenever possible, in forming programs of observation with any instrument. The observations were made in one order, and repeated in *reverse* order, thereby eliminating largely any possible progressive changes in the apparatus.

**122. Errors of graduation** exist in all circles and affect the angles measured by them. Whether these errors are negligible, or must be taken into account, depends largely upon the degree of refinement exacted by the problem in hand. In the case of small instruments constructed by first-class makers, the errors of graduation will generally be smaller than the least reading of the instrument, and may be neglected. The circles provided by the best makers for modern meridian instruments are nearly perfect. It is seldom that one of the graduations is displaced as much as $1''$ from its theoretical position, or that the mean of four graduations $90°$ apart is in error by as much as $0''.5$. Nevertheless, it is necessary to investigate every such circle to determine the degree of refinement which it will impart to observations depending upon its readings, and to secure data for eliminating errors arising from its imperfect graduation. The investigation of 10,800 graduations on a circle taxes the resources of most long established observatories so prohibitively that it is seldom or never carried to a finish. After the investigation has extended to all the graduations marking the degrees, or at the most to those marking the $20'$ divisions, the nature and magnitude of the systematic errors and the magnitude of the accidental errors will have been revealed, and further determinations may generally be confined to the graduations which are used with special frequency, *e.g.*, the graduations used in determining the nadir reading, or those used with particular stars, or zones of stars.

It will be seen from the above that the problem is one for the professional astronomer and his assistants, rather than for the student. A complete solution is therefore not called for in this place, but an outline of one of the best methods may be given.

Let us suppose that the instrument has a fixed and a movable circle, each read by four microscopes 90° apart. As an origin for the entire system of measures, let it be assumed that the mean of the readings of the four microscopes on the 0°, 90°, 180° and 270° lines is free from graduation error, in both circles. If now the axis of the instrument be rotated through a given angle, 30° for example, and the circle reading be taken, the observed angle will differ slightly from 30°, from several causes: first, the unavoidable errors of observation, which may be reduced materially by increasing the number of independent observations; second, progressive changes in the apparatus, largely due to temperature variations, which may be reduced materially by repeating the observations backwards; third, differential flexure of the circle, which may be eliminated, it is assumed, by rotating the instrument on its axis through 180° and repeating the observations on the same lines; and fourth, the graduation errors of the divisions used. These considerations suggest the principal features of the program of observations.

Let it be required, first, to determine the division errors of the 45° points of both circles; *i.e.*, the error for each circle affecting the mean of the readings obtained from the four microscopes on the points 45°, 135°, 225° and 315°. Place the 0° of the two circles in coincidence and read the microscopes for both circles. To increase the accuracy of the determination by increasing the number of observations, and at the same time eliminate the circle flexure, let these observations be repeated with the instrument rotated through one, two, three and four quadrants. Now let the 45° line of the movable circle be made coincident with the 0° line of the fixed circle, and a series of readings similar to the above be secured. Next let the 90° division of the movable circle be made coincident with the 0° line of the fixed circle, and so on until a series has been secured with each 45° division of the movable circle in coincidence with the 0° of the fixed circle. In order to eliminate progressive changes in the instrument, as far as possible, let the above program of observations be repeated in reverse order. The data will then be at hand for a thorough determination of the errors of the 45° divisions of both circles. Each arc of 45° on one circle has been compared with each such arc on the other circle. For example, the first 45° arc of the fixed circle has been used to measure each of the eight 45° arcs on the movable circle. The true sum of these eight arcs is 360°. If their sum measured by the fixed-circle arc differs from 360° by any quantity $n$, the relative division error of the mean reading of the fixed-circle microscopes on the 45° lines is $\frac{1}{8} n$. Similarly, the 45° division error for the movable circle may be computed.

The division errors of the 15° readings of the two circles may be obtained by subdividing and comparing the 45° arcs just determined, or from the complete circles, as before; and so on for the 5°, 1° and other readings.

When a circle has been investigated, the zero of the system may be changed arbitrarily by applying a constant to all the division errors secured, either to make them all of the same sign, or to make their algebraic sum zero.

The 1° readings of the fixed circle, and the 3° readings of the movable circle of the Repsold instrument of the Lick Observatory, were investigated by the above methods by R. H. Tucker.[*] The average errors of the fixed and

---

[*] For further and fuller details, see articles by Professor Tucker in *Publications Astronomical Society of the Pacific*, 1895, pp. 330–338, and 1896, pp. 270–272. Also an article by Professor Boss in *The Astronomical Journal*, 1896, Nos. 382, 383.

movable circle readings were ± 0″.18 and ± 0″.15, respectively. The following table contains the errors for the 9° readings, by way of illustration.

| Reading | Fixed Circle | Movable Circle |
|---|---|---|
| 0° | + 0″.18 | + 0″.10 |
| 9 | + 0 .12 | + 0 .14 |
| 18 | − 0 .23 | + 0 .04 |
| 27 | − 0 .04 | + 0 .37 |
| 36 | − 0 .52 | − 0 .34 |
| 45 | − 0 .34 | + 0 .03 |
| 54 | + 0 .02 | − 0 .18 |
| 63 | + 0 .11 | − 0 .07 |
| 72 | − 0 .22 | − 0 .14 |
| 81 | + 0 .16 | + 0 .08 |
| 90 | + 0 .18 | + 0 .10 |

In case the instrument has only one circle, its errors may be determined by means of two extra microscopes, placed 180° apart, in connection with the four regular microscopes.*

It should be explained that many observers shift the movable circle from time to time, so that the several observations of any star will depend upon many different graduations of the circle, thereby reducing the magnitude of the division error affecting the mean result.

The flexure of the circles in modern instruments is so small that to apply a correction for it is generally more objectionable than to omit it. This, and similar small and uncertain corrections are not ignored, however. Good practice requires that a star's position be determined from an equal number of observations with circle west and circle east, with the result that several slight errors are largely eliminated from the mean of all the observations.

---

* For an exposition of this method, see *Annalen der Sternwarte in Leiden*, Band II, seite [50–92].

THE MERIDIAN CIRCLE 187

**123.** *Reduction to the meridian.* Theoretically, the observer is supposed to bisect the image of a star with the declination micrometer at the instant of meridian passage. If for any reason the bisection is made at $t$ seconds before or after meridian passage, the necessary correction $x$ to reduce to the meridian may be found from equation (281) and the corresponding table in § 111. The correction $x$ must be applied to the circle reading with the proper sign to increase the observed polar distance.

**124.** *Refraction.* The refractions given by (95) must be applied to the circle readings in such a way as to increase the zenith distances.

**125.** *Parallax.* Observations on bodies within the solar system will require correction for parallax, by the methods of §§ 25–27.

**126.** The meridian circle is applied to the determination of declinations by two general methods.

1st. **Fundamental Determinations.** The latitude $\phi$ of the observer being known, the **equator reading** of the circle is given by

$$\text{Equator reading} = \text{Zenith reading} \pm \phi, \qquad (287)$$

the lower sign being for circle east. The difference between the circle reading for a star (corrected for refraction, etc.), and the equator reading, is the declination of the star, determined fundamentally.

2d. **Differential Determinations.** If a standard star be observed in the usual manner, its circle reading (corrected for refraction, etc.), plus or minus its known declination, will be the equator reading of the circle. The corrected circle reading for an undetermined star will differ from the equator reading by the declination of the star. Further, the equator reading obtained from the standard star, combined with the zenith reading, will furnish a new determination of the latitude.

A circumpolar star observed for latitude at both upper and lower culmination has the advantage that any error of declination is eliminated from the mean result; but the disadvantage, for observers situated well toward the equator, that the refractions are large.

Since the latitude of a place varies appreciably, fundamental determinations of declination require a knowledge of the current value of the latitude. Programs for fundamental work should contain a few standard stars, as checks on each night's results.

In programs for differential work, the undetermined stars should be preceded, accompanied and followed by standard stars; the range of declinations for the two kinds being about equal, to assist in eliminating uncertainties in refractions. The equator reading should be obtained from all the standard stars. A program covering four or five hours should contain eight or ten standard stars. If the value of the latitude is known, a long series of such observations of the standard stars will furnish corrections to the standard declinations themselves.

The following abridged program of observations, made with the Repsold meridian circle of the Lick Observatory, illustrates many of the important principles treated in this chapter.

The mean of the nadir and micrometer readings taken just before and after the observations of stars furnished the values

$$\text{Nadir reading} = 134° \, 56' \, 29''.82,$$
$$\text{Micrometer (zero reading)} = 17^r.000.$$

The meteorological observations required for computing the refractions were made at 15.2 hours sidereal time, thus:

Barom. 25.70 inches, Att. Therm. + 64° F., Ext. Therm. + 63°.0 F.

Other meteorological observations were taken throughout the program.

THURSDAY, 1898 JUNE 9. CIRCLE EAST. OBSERVER, R. H. TUCKER

| Star | γ Urs. Min. 15ᴬ 20ᵐ +72° 11′ 46″.63 | γ Libræ 15ᴬ 29ᵐ −14° 27′ 12″.61 | 5 H.Cam., L.C. 15ᴬ 39ᵐ +71° 1′ 6″.19 | Star No. 4 15ᴬ 40ᵐ −19° 5′ | Star No. 5 15ᴬ 55ᵐ −20° 52′ | β Scorpii 15ᴬ 59ᵐ −19° 31′ 48″.03 |
|---|---|---|---|---|---|---|
| Approximate α |  |  |  |  |  |  |
| δ |  |  |  |  |  |  |
| Apparent z | 34 50 46 | 51 46 36 | 71 36 4 | 56 24 22 | 58 11 25 | 56 51 0 |
| Zenith reading | 314 56 29.82 | 314 56 29.82 | 314 56 29.82 | 314 56 29.82 | 314 56 29.82 | 314 56 29.82 |
| Circle reading | 349 47 16.33 | 263 9 52.82 | 26 32 33.58 | 258 32 8.19 | 266 45 5.10 | 258 5 29.44 |
| Runs | −0.01 | +0.01 | 0.00 | 0.00 | 0.00 | +0.01 |
| Flexure | +0.06 | −0.08 | +0.09 | −0.08 | −0.08 | −0.08 |
| Refraction | +33.75 | −1 1.48 | +2 24.35 | −1 12.87 | −1 18.02 | −1 14.13 |
| Corrected circle reading | 349 47 50.13 | 263 8 51.27 | 26 34 58.02 | 258 30 55.24 | 266 43 47.00 | 258 4 15.24 |
| Observed equator reading | 277 36 3.50 | 277 36 3.88 | 277 36 4.21 | 277 36 3.58 | 277 36 3.58 | 277 36 3.27 |
| Mean equator reading |  |  |  |  |  |  |
| Observed declination | 37 20 26.32 | 37 20 25.94 | 37 20 25.01 | −19 5 8.34 | −20 52 16.58 | 37 20 26.55 |
| Observed latitude |  |  |  |  |  |  |

By way of illustration, the original circle, microscope and micrometer readings for the first star, $\gamma$ *Ursæ Minoris*, were:

| Circle | | Microscopes * | | Micrometer |
|---|---|---|---|---|
| 349° 46′ | I | 45″.1 | (48′) 45″.2 | 16ʳ.500 |
| | II | 54 .0 | 54 .0 | |
| | III | 54 .1 | 54 .0 | |
| | IV | 55 .8 | 56 .0 | |
| | Means | 52 .25 | 52 .30 | |

Original circle reading . . . . . . 349° 46′ 52″.28
Correction for micrometer, + 0ʳ.500 . +24″.05
Circle reading . . . . . . . . . 349° 47′ 16″.33

In this program, the first and third stars are standard circumpolars, the former at upper, and the latter at lower culmination. The second and sixth stars are standard stars. The declinations assigned are the apparent declinations of the standard stars, and the approximate declinations of the undetermined stars. The circle readings include the corrections for the readings of the declination micrometer. The apparent zenith distances, $z$, are the differences of the circle readings and the zenith reading, to be used in computing the refractions. The corrected circle readings, minus the declinations of the standard stars, are the observed equator readings, and their mean for the two southern stars is adopted as the equator reading. The differences of the zenith reading and the observed equator readings for the four standard stars are four values of the observed latitude. The corrected circle readings for the two undetermined stars, minus the mean equator readings are the observed declinations. The corrections for graduation errors have not been applied: the errors for the particular graduations used have not been determined.

---

* It is the custom of this observer to obtain the microscope readings on two graduations on opposite sides of the micrometer zero positions ; to use the mean of the seconds given by the eight readings ; and to correct for runs for the distance that the mean of the two graduations is from the micrometer zero. The value of the correction for runs is determined from all the readings of the program.

# CHAPTER X

## ASTRONOMICAL AZIMUTH

**127.** In many problems in higher surveying the azimuth of a point on the earth's surface, as viewed from the point of observation, is required to be very accurately known. It is determined by measuring the *difference* of the azimuths of the point and a star by means of a theodolite, a surveyor's transit, or any similar instrument designed for measuring horizontal angles. The azimuth of the star is computed from the known right ascension, declination, latitude and time; whence the azimuth of the point can be obtained.

Only the four circumpolar stars, whose places are given in the Ephemeris, pp. 302–313, should be used in accurate determinations.

The point whose azimuth is to be determined is marked conveniently by a lamp arranged to shine through a small hole in a box, placed directly over the point. It should be at least one mile from the observer. If no provision has been made for illuminating the wires of the telescope at night, they can be rendered visible by tying a piece of thin unglazed white paper over the object glass of the telescope, first cutting a hole in the paper nearly as large as the object glass, and throwing the light of a bull's-eye lantern on the paper.

The instrument is set up over the point of observation (marked in some way) and carefully adjusted. The horizontal graduated circle is *fixed* in position by clamping. The level screws and other adjusting screws must not be touched *during* a series of observations. If the rotation axis of the telescope is not truly horizontal, an error is introduced in the measured difference of azimuth of the mark and star, which must be allowed for. In the finer instruments the inclination of the axis is measured by means of a striding level. The effect of an error of collimation is practically eliminated by reversing the instrument and observing an equal number of times in both positions. We shall consider that this is always done. If more than one series of observations is made, the horizontal circle should be shifted so that a different part of it may be used, thereby eliminating largely any errors of graduation of the circle.

**128.** *Correction for Level.* When the rotation axis of the telescope is inclined to the horizon, the line of sight does not describe a vertical circle, and the horizontal circle reading requires a small correction. Let $b$ be the elevation of the west end of the axis above the horizon, and let the west end of the produced axis cut the celestial sphere in $W$; let $y$ be the corresponding correction to the circle reading; and let $Z$ be the zenith and $S$ the star. Then, in the triangle $WZS$,

$$WZ = 90° - b, \quad ZS = z, \quad WS = 90°, \quad WZS = 90° + y;$$

and, therefore,

$$\sin b \cos z - \cos b \sin z \sin y = 0.$$

But $b$ and $y$ are very small, and we may write

$$y = b \cot z; \tag{288}$$

and for circumpolar stars, we may write

$$y = b \tan \phi. \tag{289}$$

The value of $b$ is found by (166) or (167), or by the methods of § 94. If the illuminated mark is not in the horizon, the circle readings on the mark must be corrected by (288), using its zenith distance $z$.

**129.** *Correction for diurnal aberration.* Owing to diurnal aberration the star will be observed too far east. In the most refined observations this must be allowed for. The correction to the circle reading is given by (118); which, for circumpolar stars, is approximately

$$dA = + 0''.31 \cos A. \tag{290}$$

If the circles cannot be read to less than $1''$, this correction is negligible.

**130.** *Correction for error of runs.* If reading microscopes are used [§ 58], the circle readings must be further corrected for error of runs.

### AZIMUTH BY A CIRCUMPOLAR STAR NEAR ELONGATION

**131.** A star is at **western** or **eastern elongation** when its azimuth is the least or greatest possible. It is then moving in a vertical circle, and is in the most favorable position for azimuth observations. Only one observation can be made *at* the instant of elongation, but it is customary to make several observations just before and after elongation, and allow for the change in azimuth during the intervals.

At the instant of elongation, the triangle formed by the pole, star and zenith, which we shall denote by $PSZ$, is right-angled at the star. If we let $\theta_0$ be the sidereal time, and $A_0$, $t_0$ and $z_0$ the azimuth, hour angle and zenith distance of the star at elongation, $a$ and $\delta$ its right ascension and declination, and $\phi$ the observer's latitude, we shall have, for western elongation,

$$PZ = 90° - \phi, \quad PS = 90° - \delta, \quad ZS = z_0,$$
$$PZS = 180° - A_0, \quad ZPS = t_0, \quad PSZ = 90°;$$

and for eastern elongation,

$$PZS = A_0 - 180°, \quad ZPS = 360° - t_0.$$

o

We can write

$$\cos t_0 = \frac{\tan \phi}{\tan \delta}, \quad \cos z_0 = \frac{\sin \phi}{\sin \delta}, \quad \sin A_0 = \pm \frac{\cos \delta}{\cos \phi}, \quad (291)$$

$$\theta_0 = a + t_0. \quad (292)$$

$t_0$ is in the first quadrant for western elongation, and in the fourth for eastern; $z_0$ is always in the first quadrant; and $A_0$ is less than 180° for western elongation, and greater than 180° for eastern. We can also write

$$\pm \sin A_0 = \frac{\cos \delta}{\cos \phi} = \frac{\sin \delta \cos t_0}{\sin \phi}, \quad (293)$$

$$\pm \cos A_0 = -\sin \delta \sin t_0, \quad (294)$$

in which the upper signs are for western elongation, the lower for eastern.

If the star is observed at any other hour angle $t$, its azimuth $A$ is given by (16) and (17). Multiplying (16) by (293), (17) by (294), and subtracting one product from the other, we obtain

$$\sin z \sin (A_0 - A) = \mp \sin \delta \cos \delta \, 2 \sin^2 \tfrac{1}{2} (t_0 - t). \quad (295)$$

If the observations are made near elongation, $t$ will not differ much from $t_0$, $A_0 - A$ will be small, and for the circumpolar stars $z$ will not differ much from $z_0$. Therefore we can write, without sensible error,

$$A_0 - A = \mp \frac{\sin \delta \cos \delta}{\sin z_0} \cdot \frac{2 \sin^2 \tfrac{1}{2} (t_0 - t)}{\sin 1''}, \quad (296)$$

in which the lower sign is for eastern elongation, as before. $A_0 - A$ is the correction to be applied to the circle reading for an observation made at hour angle $t$, to reduce to the corresponding circle reading for an observation made at hour angle $t_0$.

For convenience, let

$$m = \frac{2 \sin^2 \tfrac{1}{2} (t_0 - t)}{\sin 1''},$$

and (296) becomes

$$A_0 - A = \mp m \frac{\sin \delta \cos \delta}{\sin z_0}. \quad (297)$$

DETERMINATION OF AZIMUTH        195

The values of $m$ can be taken from Table III, Appendix, for the different values of $t_0 - t$. If we let $m_0$ be the mean of the several values of $m$, the corrections can be applied collectively to the mean of the circle readings on the star, and the equation (297) becomes

$$A_0 - A = \mp m_0 \frac{\sin \delta \cos \delta}{\sin z_0}, \qquad (298)$$

in which $A_0 - A$ is the correction to the mean of the circle readings. Further, if the level readings have been taken symmetrically with reference to the program, which can always be done, the mean value of $y$, equation (289), can be applied to the mean of the circle readings.

**132.** The values of $t_0$ and $\theta_0$ having been computed for the star to be observed, the instrument is carefully adjusted, and a program similar to this is followed:

>Make two readings on the mark
>Read the level
>Make four readings on the star
>Read the level
>Make two readings on the mark
>Reverse
>Make two readings on the mark
>Read the level
>Make four readings on the star
>Read the level
>Make two readings on the mark

The times of observation are noted on a time-piece, preferably a sidereal chronometer. Its correction must be known within one or two seconds if the most refined form of instrument is employed, or to the nearest minute if an ordinary surveyor's transit is used. This correction can be obtained by any of the methods described in the preceding chapters, or by a comparison with the time signals at the nearest telegraph station. The chronometer time of elongation is now known. Subtracting from it the several times of observation, the values of $t_0 - t$ are found, and the values of $m$ corresponding to them taken from Table III. Forming the mean $m_0$ and computing $z_0$ from (291), the value of $A_0 - A$ is found and applied to the mean of the circle readings. The mean of the corrections for level errors and the correction for diurnal aberration are now applied. The corrected mean circle reading, which we shall call $s$, corresponds to the azimuth $A_0$ of the star at elongation, which is computed by (291). If $k$ is the mean of the circle readings on the mark, and $M$ the azimuth of the mark, then

$$M = k - (s - A_0). \tag{299}$$

The circle reading when the telescope is directed to the south point of the horizon will be equal to $s - A_0$.

**133. Example.** Detroit Observatory, Wednesday, 1891 May 6. Find the azimuth of a given point (nearly in the horizon) from observations on $\delta$ *Ursæ Minoris* near its eastern elongation. Observer's latitude, $42° 16' 48''$.

The apparent place of the star was

$$\alpha = 18^h 7^m 44^s, \qquad \delta = + 86° 36' 25''.0.$$

Equations (291) and (292) are solved as below.

| | | |
|---|---|---|
| $\tan \phi$ 9.958704 | $\sin \phi$ 9.827856 | $\cos \delta$ 8.772214 |
| $\tan \delta$ 1.227024 | $\sin \delta$ 9.999236 | $\cos \phi$ 9.869153 |
| $t_0$ 273° 5′ 25″ | $z_0$ 47° 37′ 42″ | $A_0$ 184° 35′ 17″.8 |
| $t_0$ 18$^h$ 12$^m$ 22$^s$ | | |
| $\alpha$ 18  7  44 | | |
| $\theta_0$ 12 20  6 | | |

The chronometer correction was $+ 18^m 52^s$, and, therefore, the chronometer time of elongation was $12^h 1^m 14^s$. A good surveyor's transit, whose horizontal circle was read to $10''$ by two verniers, and which was provided with plate levels and a delicate striding level, was placed over the point of observation and carefully leveled a short time before elongation. The following observations were made:

DETERMINATION OF AZIMUTH          197

| No. | Object | Telescope | Chronometer | Vernier A | Vernier B |
|---|---|---|---|---|---|
| (1) | Mark | Reversed |  | 96° 16′ 40″ | 276° 16′ 30″ |
| (2) | " | " |  | 96 16 35 | 276 16 25 |
| (3) | Level |  |  |  |  |
| (4) | Star | " | 11ʰ 44ᵐ 52ˢ | 243 39 20 | 63 39 20 |
| (5) | " | " | 48 40 | 243 39 50 | 63 39 50 |
| (6) | " | " | 51 6 | 243 39 50 | 63 39 50 |
| (7) | " | " | 53 11 | 243 40 0 | 63 40 5 |
| (8) | Level |  |  |  |  |
| (9) | Mark | " |  | 96 16 45 | 276 16 40 |
| (10) | " | " |  | 96 16 50 | 276 16 40 |
| (11) | " | Direct |  | 276 17 0 | 96 16 45 |
| (12) | " | " |  | 276 16 55 | 96 16 45 |
| (13) | Level |  |  |  |  |
| (14) | Star | " | 12 5 50 | 63 40 10 | 243 40 10 |
| (15) | " | " | 7 54 | 63 40 0 | 243 40 0 |
| (16) | " | " | 9 44 | 63 39 50 | 243 39 50 |
| (17) | " | " | 11 27 | 63 39 45 | 243 39 55 |
| (18) | Level |  |  |  |  |
| (19) | Mark | " |  | 276 16 45 | 96 16 30 |
| (20) | " | " |  | 276 16 45 | 96 16 35 |

The level readings given by the striding level were

| (3) | | (8) | | (13) | | (18) | |
|---|---|---|---|---|---|---|---|
| W | E | W | E | W | E | W | E |
| 4.4 | 4.1 | 4.2 | 4.0 | 4.0 | 4.4 | 4.0 | 4.2 |
| 4.3 | 4.0 | 4.3 | 3.8 | 4.0 | 4.2 | 4.1 | 4.2 |
| 4.4 | 4.0 | 4.2 | 3.9 | 4.0 | 4.2 | 4.0 | 4.2 |
| 4.3 | 3.8 | 4.4 | 3.8 | 4.1 | 4.4 | 4.2 | 4.2 |

The value of one division of the level was $10''.7$, and therefore, from (166), the inequality of the pivots being negligible,

$$\begin{array}{cccc} (3) & (8) & (13) & (18) \\ b = +2''.0, & +2''.1, & -1''.4, & -0''.6, \end{array}$$

and by (289),

$$\begin{array}{cccc} (3) & (8) & (13) & (18) \\ y = +1''.8, & +1''.9, & -1''.3, & -0''.6. \end{array}$$

The solution of (298) for the eight readings on the star is given below. The column "Circle Readings" is formed by taking the means of Vernier A and Vernier B.

| No. | Circle Readings | $t_0 - t$ | $m$ |
|---|---|---|---|
| (4)  | 243° 39' 20" | $+ 16^m\ 22^s$ | 525".7 |
| (5)  | 243  39  50  | $+ 12\ \ 34$    | 310 .0 |
| (6)  | 243  39  50  | $+ 10\ \ \ 8$   | 201 .6 |
| (7)  | 243  40  2   | $+\ \ 8\ \ \ 3$ | 127 .2 |
| (14) | 63   40  10  | $-\ \ 4\ \ 36$  | 41 .5 |
| (15) | 63   40  0   | $-\ \ 6\ \ 40$  | 87 .3 |
| (16) | 63   39  50  | $-\ \ 8\ \ 30$  | 141 .8 |
| (17) | 63   39  50  | $- 10\ \ 13$    | 204 .9 |
| Means | 63  39  51.5 |                 | $m_0 = 205\ .0$ |

$$\begin{aligned}
\log m_0 &\quad 2.31175 \\
\sin \delta &\quad 9.99924 \\
\cos \delta &\quad 8.77221 \\
\operatorname{cosec} z_0 &\quad 0.13148 \\
\log (A_0 - A) &\quad 1.21468 \\
A_0 - A &\quad + 16".4
\end{aligned}$$

The mean of the four values of $y$ is $+ 0".4$. The value of $dA$, from (290), is $- 0".3$. The corrected circle reading on the star at elongation is therefore

$$s = 63°\ 39'\ 51".5 + 16".4 + 0".4 - 0".3 = 63°\ 40'\ 8".0.$$

The mean of all the readings on the mark is

$$k = 276°\ 16'\ 41".6;$$

and, therefore, by (299),

$$M = 37°\ 11'\ 51".4.$$

Since the verniers on this instrument read to only 10", the diurnal aberration could have been neglected, and the other corrections computed to the nearest second only. But all the corrections have been applied here, to illustrate the method.

DETERMINATION OF AZIMUTH 199

AZIMUTH BY POLARIS OBSERVED AT ANY HOUR ANGLE

**134.** When the azimuth is required with the greatest possible accuracy, the observations should always be made at or near elongation, and reduced as in the preceding section. However, good results can be obtained by observing *Polaris* in any position, if the time is accurately known. The time should be known within $0^s.5$ when using the finest instruments, and within $5^s$ or $10^s$ when using a good surveyor's transit whose least reading is $10''$.

As in the preceding method, the observations should be made on the mark and star in both positions of the telescope. If the observations are made in quick succession, the mean of two or three observations made before reversing may be treated as a single observation, and similarly for those made after reversing. But if the separate observations are several minutes apart, each observation should be reduced separately.

The sidereal time $\theta$ of observation having been noted with great care, the hour angle $t$ of *Polaris* is given by (41). If we let the azimuth $A$ of the star be measured *from the north point*, $+$ if the star is west of the meridian and $-$ if east, and let $q$ be the star's parallactic angle [§ 6], we may write [*Chauvenet's Sph. Trig.*, § 24]

$$\tan \tfrac{1}{2}(q+A) = \frac{\cos \tfrac{1}{2}(\delta - \phi)}{\sin \tfrac{1}{2}(\delta + \phi)} \cot \tfrac{1}{2} t = f \cot \tfrac{1}{2} t, \qquad (300)$$

$$\tan \tfrac{1}{2}(q-A) = \frac{\sin \tfrac{1}{2}(\delta - \phi)}{\cos \tfrac{1}{2}(\delta + \phi)} \cot \tfrac{1}{2} t = f' \cot \tfrac{1}{2} t, \qquad (301)$$

$$A = \tfrac{1}{2}(q+A) - \tfrac{1}{2}(q-A). \qquad (302)$$

The auxiliary quantities, $f$ and $f'$, depending on $\delta$ and $\phi$, are constant for a night's observations, and with surveyors' instruments may be considered constant for several weeks. When they have been computed, once for the whole series of observations, they may be combined rapidly with the values of $\cot \frac{1}{2} t$ for the individual observations, to determine $\frac{1}{2}(q + A)$ and $\frac{1}{2}(q - A)$, and thence $A$ by (302). The values of $q$ need not be determined at all. The correction for level is given, as before, by (288), and for diurnal aberration by (290).

With $A$ representing the azimuth of the star measured from the north point as defined above, and computed by means of (300), (301) and (302), let $S$ be the circle reading on the star corrected for level and aberration, $K$ the mean of all the readings on the mark, and $N$ the azimuth of the mark measured from the north point, + if west of north, and − if east. Then, assuming that the circle readings increase in the direction of motion of the hands of a watch, we shall have

$$N = S + A - K; \qquad (303)$$

and $S + A$ will be the circle reading when the instrument points due north.

*Example.* Detroit Observatory, Wednesday, 1891 May 6. Find the azimuth of a given point nearly in the horizon, from the following observations of *Polaris*, made with the instrument described in § 133.

| No. | Object | Telescope | Chronometer | Vernier A | Vernier B |
|---|---|---|---|---|---|
| (1) | Mark | Direct |  | 276° 16′ 40″ | 96° 16′ 35″ |
| (2) | " | " |  | 276 16 40 | 96 16 30 |
| (3) | Level |  |  |  |  |
| (4) | Star | " | 13ʰ 22ᵐ 59ˢ | 59 15 50 | 239 15 40 |
| (5) | " | " | 13 26 30 | 59 17 0 | 239 16 50 |
| (6) | " | " | 13 27 57 | 59 17 30 | 239 17 20 |
| (7) | " | Reversed | 13 32 47 | 239 19 40 | 59 19 45 |
| (8) | " | " | 13 34 56 | 239 20 35 | 59 20 30 |
| (9) | " | " | 13 36 40 | 239 21 20 | 59 21 15 |
| (10) | Level |  |  |  |  |
| (11) | Mark | " |  | 96 16 40 | 276 16 30 |
| (12) | " | " |  | 96 16 40 | 276 16 35 |

DETERMINATION OF AZIMUTH 201

The striding level gave

| (3) | | (10) | |
|---|---|---|---|
| W | E | W | E |
| 3.6 | 5.3 | 4.9 | 4.1 |
| 4.6 | 4.3 | 4.5 | 4.5 |

whence, by (166),

(3)           (10)
$b = -3''.3,$   $+2''.1.$

The position of *Polaris* was, American Ephemeris, p. 306,

$$a = 1^h\ 17^m\ 53^s, \qquad \delta = 88°\ 43'\ 29'',$$

and the chronometer correction was $+18^m\ 52^s$.

The means of the observations made before and after reversal are reduced below.

| | Before | After |
|---|---|---|
| Chronometer | $13^h\ 25^m\ 49^s$ | $13^h\ 34^m\ 48^s$ |
| $\Delta\theta$ | $+18\ \ 52$ | $+18\ \ 52$ |
| $\theta$ | $13\ 44\ 41$ | $13\ 53\ 40$ |
| $a$ | $1\ 17\ 53$ | $1\ 17\ 53$ |
| $\theta - a = t$ | $12\ 26\ 48$ | $12\ 35\ 47$ |
| $t$ | $186°\ 42'\ 0''$ | $188°\ 56'\ 45''$ |
| $\tfrac{1}{2}t$ | $93\ 21\ \ 0$ | $94\ 28\ 22$ |
| $\delta$ | $+\ 88\ 43\ 29$ | |
| $\phi$ | $+\ 42\ 16\ 48$ | |
| $\delta - \phi$ | $+\ 46\ 26\ 41$ | |
| $\delta + \phi$ | $+131\ \ 0\ 17$ | |
| $\tfrac{1}{2}(\delta - \phi)$ | $+\ 23\ 13\ 20$ | |
| $\tfrac{1}{2}(\delta + \phi)$ | $+\ 65\ 30\ \ 8$ | |
| $\cos\tfrac{1}{2}(\delta - \phi)$ | $9.963307$ | |
| $\sin\tfrac{1}{2}(\delta + \phi)$ | $9.959031$ | |
| $\log f$ | $0.004276$ | $0.004276$ |
| $\cot\tfrac{1}{2}t$ | $8.767417_n$ | $8.893338_n$ |
| $\tfrac{1}{2}(q + A)$ | $-3°\ 22'\ 59''$ | $-4°\ 31'\ 1''$ |
| $\sin\tfrac{1}{2}(\delta - \phi)$ | $9.595825$ | |
| $\cos\tfrac{1}{2}(\delta + \phi)$ | $9.617690$ | |
| $\log f'$ | $9.978135$ | $9.978135$ |
| $\cot\tfrac{1}{2}t$ | $8.767417_n$ | $8.893338_n$ |

## 202 PRACTICAL ASTRONOMY

| | | |
|---|---|---|
| $\frac{1}{2}(q-A)$ | $-3°\ 11'\ \ 9''$ | $-4°\ 15'\ 14''$ |
| $A$ | $-0\ \ 11\ 50$ | $-0\ \ 15\ 47$ |
| Circle on star | $59°\ 16'\ 42''$ | $239°\ 20'\ 31''$ |
| $y$ | $-3$ | $+2$ |
| $dA$ | $0$ | $0$ |
| $S$ | $59\ \ 16\ 39$ | $239\ \ 20\ 33$ |
| $S + A$ | $59\ \ \ 4\ 49$ | $239\ \ \ 4\ 46$ |
| $K$ | $276\ \ 16\ 36$ | $96\ \ 16\ 36$ |
| $N$ | $142\ \ 48\ 13$ | $142\ \ 48\ 10$ |
| Mean $N$ | | $142°\ 48'\ 12''$ |

The corrections should be carried to tenths of seconds when refined instruments are used.

The azimuth of the same mark was measured on the same night with the same instrument, by means of observations of $\delta$ *Ursæ Minoris* taken near eastern elongation [see example of the preceding section]. The azimuth obtained, measured from the south point, was

$$M = 37°\ 11'\ 51''.4.$$

# CHAPTER XI

## THE SURVEYOR'S TRANSIT

**135.** The surveyor's transit is adapted to the determination of the time, latitude and azimuth by many of the preceding methods. These elements can easily be determined to an accuracy within the least readings of the circles, if the instrument is of reliable make, and is provided with spirit levels. We shall assume that the observer uses a *mean* time-piece, which we shall call a watch, and *that he has a thorough knowledge of the subject of* TIME, CHAPTER II, *without which the Ephemeris cannot be used intelligently*. We shall assume, also, that the vertical circle of his instrument is complete, and that the degrees are numbered consecutively from 0 to 360. In case they are not, the observer can readily reduce his readings to that system. The instrument is supposed to be carefully adjusted. A method of illuminating the wires at night is given in § 127.

Figure 26 illustrates a form of instrument well adapted to the solution of the problems described in this chapter; but the methods can be used, within limits, with nearly all forms of the surveyor's transit instrument.

### DETERMINATION OF TIME

**136.** *By equal altitudes of a star.* Set the instrument up firmly, level it, and direct the telescope to a known bright star east of the meridian. Pointing the telescope slightly above the star, clamp the vertical circle and note the time $T'$ when the star crosses the horizontal wire.

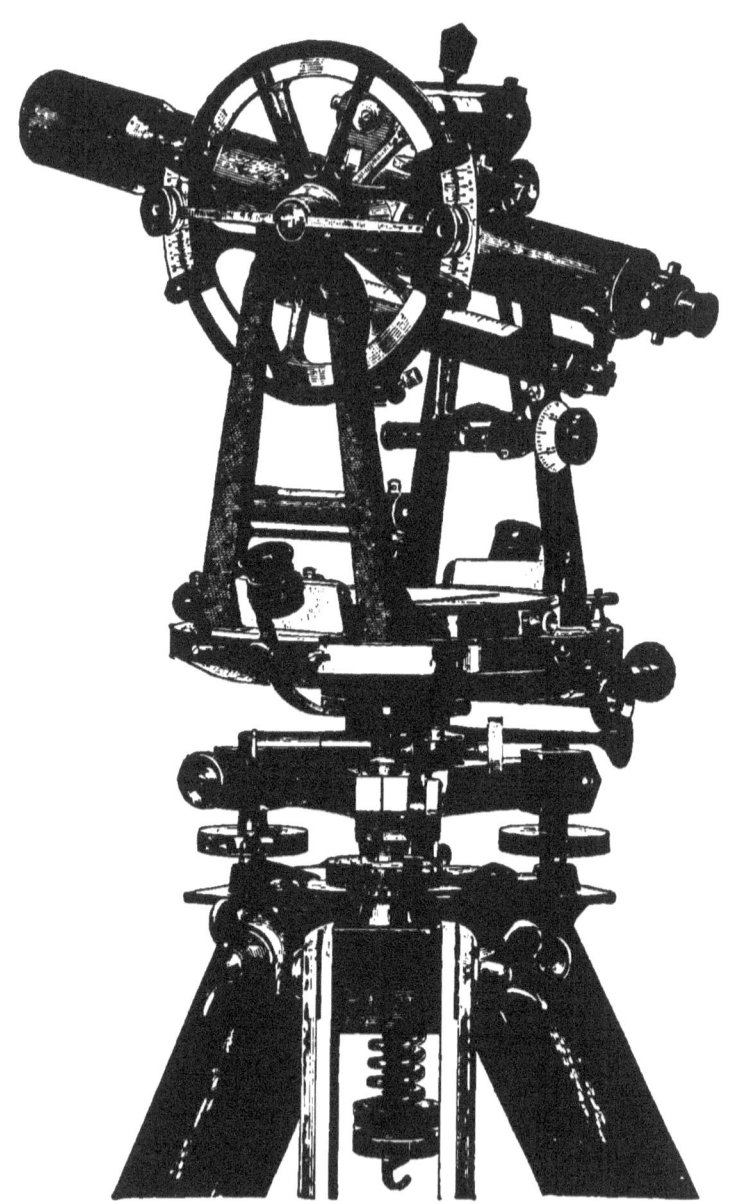

FIG. 26

The vertical circle must not be unclamped. A short time before the star reaches the same altitude west of the meridian, level the instrument, move it in azimuth until the telescope is directed to a point just below the star, wait for the star to enter the field, and note the time $T''$ when it crosses the horizontal wire. The sidereal time $\theta$ when the star was on the observer's meridian equals its right ascension $a$, and this corresponds to the mean of the two watch times. Converting the sidereal time $\theta = a$ into the corresponding mean time $T$, the watch correction $\Delta T$ is given by

$$\Delta T = T - \tfrac{1}{2}(T' + T''). \qquad (304)$$

*Example.* Thursday, 1891 March 5. In longitude $5^h\ 34^m\ 55^s$, *Regulus* was observed at equal altitudes east and west of the meridian, at the watch times

$$T' = 8^h\ 7^m\ 34^s, \qquad T'' = 14^h\ 10^m\ 20^s.$$

Required the watch correction.

From the American Ephemeris, p. 332, $a = \theta = 10^h\ 2^m\ 35^s$. Converting this into mean time, § 18, we find

$$T = 11^h\ 8^m\ 3^s;$$

and, therefore, by (304), the watch correction was

$$\Delta T = -54^s;$$

or the watch was $54^s$ fast.

**137.** *By a single altitude of a star.* Level the transit instrument. Direct the telescope very slightly above a known star in the east or below a known star in the west, and clamp the telescope. Note the watch time when the star crosses the horizontal wire and read the vertical circle. Unclamp the telescope and repeat the observation once or twice, as quickly as possible. Double reverse the instrument, and make the *same number* of observations as before. Form the means of the circle readings made before reversal and of those made after. Subtract one from the other in that order which makes their difference less than 180°. One half this difference is the apparent zenith distance of the star at $T'$, the mean of the several watch times of observation. Adding the refraction given by (97),

$$r = 58'' \tan z, \qquad (305)$$

the result is the true zenith distance $z$. Substituting the values of $z$, $\phi$ and $\delta$ in (38) or (39), the hour angle $t$ is found. The sidereal time $\theta$ is given by [§ 8],

$$\theta = \alpha + t. \qquad (306)$$

Converting $\theta$ into the mean time $T$, the watch correction is given by

$$\Delta T = T - T'. \qquad (307)$$

*Example.* Saturday, 1891 April 25. In latitude $+ 42° 16' 47''$ and longitude $5^h 34^m 55^s$, the following altitudes of *Arcturus* were observed east of the meridian. Find the watch correction.

| Telescope | Watch | Circle reading |
|---|---|---|
| Direct | $7^h 52^m 23^s$ | 34° 43′ 0″ |
| " | 53 33 | 34 55 30 |
| " | 54 20 | 35 4 0 |
| Reversed | 55 37 | 144 42 30 |
| " | 56 30 | 144 33 0 |
| " | 57 18 | 144 24 0 |

The means of the circle readings are 34° 54′ 10″ and 144° 33′ 10″; and one half their difference is the

| | | | |
|---|---|---|---|
| Apparent zenith distance, | 54° 49′ 30″ | log 58 | 1.7634 |
| Refraction, $r$, | 1 22 | tan $z$ | 0.1520 |
| True zenith distance, $z$, | 54 50 52 | log $r$ | 1.9154 |
| | | $r$ | 82″ |

From the Ephemeris, p. 340,

$$\alpha = 14^h 10^m 43^s, \qquad \delta = + 19° 44' 52''.$$

DETERMINATION OF TIME 207

The solution of (38) is

| | | | |
|---|---|---|---|
| $z$ | 54° 50′ 52″ | $\log \sin \tfrac{1}{2}[z+(\phi-\delta)]$ | 9.79595 |
| $\phi$ | + 42 16 47 | $\log \sin \tfrac{1}{2}[z-(\phi-\delta)]$ | 9.44449 |
| $\delta$ | + 19 44 52 | $\log \sec \tfrac{1}{2}[z+(\phi+\delta)]$ | 0.28114 |
| $\phi-\delta$ | 22 31 55 | $\log \sec \tfrac{1}{2}[z-(\phi+\delta)]$ | 0.00085 |
| $\phi+\delta$ | 62 1 39 | $\log \tan^2 \tfrac{1}{2}t$ | 9.52243 |
| $z+(\phi-\delta)$ | 77 22 47 | $\log \tan \tfrac{1}{2}t$ | 9.76121$_n$ |
| $z-(\phi-\delta)$ | 32 18 57 | $\tfrac{1}{2}t$ | 150° 0′ 48″ |
| $z+(\phi+\delta)$ | 116 52 31 | $t$ | 300 1 36 |
| $z-(\phi+\delta)$ | − 7 10 47 | $t$ | 20$^h$ 0$^m$ 6$^s$ |

Solving (306), $\theta = 10^h 10^m 49^s$. The equivalent mean time is $T = 7^h 55^m 45^s$; and the mean of the six times of observation is $T' = 7^h 55^m 2^s$. Therefore,

$$\Delta T = + 43^s.$$

**138.** *By a single altitude of the sun.**  Observe the transits of the sun's upper and lower limbs over the horizontal wire by the method used for a star, § 137. Double reverse, and repeat the observations.† Form half the difference of the means of the circle readings, and add the refraction given by (305), as before. Further, subtract the parallax given by (64)

$$p = 9'' \sin z, \qquad (308)$$

and the result is the sun's true zenith distance $z$ at the mean of the times, $T'$. The correct mean time is probably known within 5$^m$ or 10$^m$. Increase it by the longitude, and the result is an approximate value of the Greenwich mean time. Take from the Ephemeris, p. II of the month, the value of the sun's declination $\delta$ at that time. The

---

* The observer must cover the eyepiece with a small piece of very dense neutral-tint glass before looking through at the sun. The observations can be made, also, by holding a piece of paper a short distance back from the eyepiece, and focusing the eyepiece so that the images of the sun and wire are seen on the paper.

† While waiting for the second limb to approach the wire, the time may well be spent in reading the vertical circle.

Ephemeris contains the apparent declination for Greenwich mean noon, and the " difference for one hour," whence the declination at any instant can be found. Solve (38) or (39) for these values of $z$, $\delta$ and $\phi$. The resulting hour angle $t$ is the observer's *apparent solar time*. Convert this into the equivalent mean time $T$, by § 15. The watch correction is given by (307), as before.

*Example.* Thursday morning, 1891 May 6. In latitude $+ 42° 16' 50''$ and longitude $5^h 35^m$, the following observations of the sun were made with Buff & Berger transit No. 1554. Required the watch correction.

| Telescope | Limb | Watch | Circle reading |
|---|---|---|---|
| Direct | Upper | $20^h 38^m 59^s$ | 41° 48' 30'' |
| " | Lower | 41 59 | 41 48 30 |
| Reversed | Upper | 46 49 | 137 33 0 |
| " | Lower | 49 50 | 137 33 0 |

One half the difference of the circle readings is $47° 52' 15''$. The refraction, by (305), is $64''$, and the parallax, by (308), is $7''$. Therefore the true zenith distance $z$ of the sun's center is $47° 53' 12''$. The mean of the four watch times is $20^h 44^m 24^s$. We have

| | | |
|---|---|---|
| $T'$ | 1891 May | $6^d 20^h 44^m 24^s$ |
| Longitude | | 5 35 |
| Gr. mean time | | 7 2 19 |
| " " " | | 7 $2^h.32$ |

From the American Ephemeris, p. 75, the sun's declination at Greenwich mean noon, May 7, was $+ 16° 48' 53''$, and the difference for one hour, $+ 41''$. The change for $2^h.32$ was therefore $96''$, and the required value of the declination was $\delta = + 16° 50' 29''$.

Substituting the values of $z$, $\phi$ and $\delta$ in (38), and solving as was done in § 137, we obtain the hour angle $t = 312° 12' 10'' = 20^h 48^m 49^s$. The observer's true time is therefore May $6^d 20^h 48^m 49^s$. Converting this into the

DETERMINATION OF LATITUDE     209

mean time $T$, by § 15, we find $T = 1891$ May $6^d\,20^h\,45^m\,15^s$.
The watch correction is

$$\Delta T = 20\ 45^m\,15^s - 20^h\,44^m\,24^s = +\,51^s.$$

## DETERMINATION OF GEOGRAPHICAL LATITUDE

**139.** *By a meridian altitude of a star.* A star is on the observer's meridian when the sidereal time $\theta$ is equal to its right ascension $a$. Convert this into the corresponding mean time, subtract the watch correction obtained by any of the above methods from it, and the result is the watch time of the star's meridian passage. A few seconds before this watch time direct the telescope to the star, bring the star's image on the horizontal wire, and read the circle. Double reverse quickly, and make another observation. As before, form one-half the difference of the circle readings, add the refraction given by (305), and the sum is the star's true zenith distance $z$. Take the value of $\delta$ from the Ephemeris. For a star observed south of the zenith,

$$\phi = \delta + z;\qquad(309)$$

and for a star observed between the zenith and pole,

$$\phi = \delta - z.\qquad(310)$$

For a star below the pole the sidereal time of meridian passage is $12^h + a$. Obtaining the value of $z$ as before, the latitude is given by

$$\phi = 180° - \delta - z.\qquad(311)$$

*Example.* Ann Arbor, Friday, 1891 April 24. $a$ *Hydræ* was observed on the meridian with a surveyor's transit, as below. Required the latitude.

| Telescope | Circle reading |
|---|---|
| Reversed | 140° 26′ 30″ |
| Direct | 39  33   0 |

One half the difference of the circle readings is 50° 26′ 45″. The refraction is 70″. Therefore, $z = 50°\,27'\,55''$.

P

From the Ephemeris, p. 331, $\delta = -8° 11' 18''$. Therefore, from (309),

$$\phi = 42° 16' 37''.$$

To find the watch time when the star is on the meridian, we have, from the Ephemeris, $a = \theta = 9^h 22^m 14^s$. The corresponding mean time, by § 18, is $7^h 11^m 13^s$. The watch correction is $+ 43^s$, whence the required watch time is $7^h 10^m 30^s$.

**140.** *By a meridian altitude of the sun.* The sun is on the meridian at the apparent time $0^h 0^m 0^s$. Apply the equation of time to this, by § 15, and subtract the known watch correction. The result is the watch time of the sun's meridian passage. One or two minutes before this watch time, direct the horizontal wire of the telescope to the upper limb of the sun, and read the vertical circle. Observe the lower limb in the same way. Double reverse and observe both limbs again. Form half the difference of the means of the readings in the two positions. Add the refraction given by (305), and subtract the parallax given by (308). The result is the value of $z$. Take from the Ephemeris the value of $\delta$ for the time of meridian passage. The latitude is now given by (309), as in the case of a star.

*Example.* Wednesday, 1891 March 25. In longitude $5^h 35^m$ the following meridian altitude observations of the sun were made with a surveyor's transit. Required the latitude.

| Telescope | Limb | Circle reading |
|---|---|---|
| Direct | Upper | 49° 54' 30'' |
| " | Lower | 49 22 30 |
| Reversed | " | 130 5 30 |
| " | Upper | 130 38 30 |

One half the difference of the means of the circle readings is $40° 21' 45''$. The refraction is $49''$. The parallax is $6''$. Therefore, $z = 40° 22' 28''$.

DETERMINATION OF AZIMUTH 211

The Greenwich apparent time of observation was March
$25^d\ 5^h\ 35^m$. The value of $\delta$ at that instant was $+ 1° 54' 32''$,
Ephemeris, p. 38. Therefore, by (309),

$$\phi = 42° 17' 0''.$$

To find the watch time of meridian passage, we have,

Apparent time   $0^h\ 0^m\ 0^s$
Equation of time  $+ 6\ \ 3$
Mean time    $0\ 6\ \ 3$
Watch correction  $- 0\ 15$
Watch time    $0\ 6\ 18$

## DETERMINATION OF AZIMUTH

**141.** The two methods of determining azimuth which
are described in the preceding chapter are adapted to the
surveyor's transit, and need no further explanation. With
this instrument the diurnal aberration can be neglected.

If the transit is provided with plate levels only, they
should be kept in perfect adjustment. If the bubble of
the level which is parallel to the rotation axis of the telescope remains constantly in the center, no correction for
level is required. But if the bubble is $n$ divisions of
the level from the center when an observation on a star
is made, and $d$ is the value of one division of the level,
the circle reading must be corrected by

$$y = \pm\ nd \cot z; \qquad (312)$$

$+$ if the bubble is too far west, $-$ if too far east.

## CHAPTER XII

### THE EQUATORIAL

**142.** This instrument consists essentially of the following parts: A supporting **pier**; a **polar axis** parallel to the earth's axis, supported at two or more points by the pier in such a way that it can rotate; a **declination axis** attached to the upper end of the polar axis, and at right angles to it, in such a way that it can rotate; a **telescope** firmly attached to one end of the declination axis, and at right angles to it; a graduated **declination circle** attached to the other end of the declination axis; a graduated **hour circle** attached to the polar axis, and at right angles to it; a finding telescope or **finder**, to assist in pointing the principal telescope, and attached to it; a **driving clock** and train of wheels for rotating the instrument about its polar axis at a uniform rate. The various moving parts are so counterpoised that the telescope will be in equilibrium in all positions. The principal features of the equatorial are well illustrated in Fig. 27.

A sidereal chronometer is an almost indispensable companion of the equatorial.

The equatorial serves two purposes:

1st. As an instrument of direct observation and discovery, by assisting the vision.

2d. As an instrument for determining very accurately the relative positions of two objects comparatively near each other, by means of a micrometer eyepiece [§ 60]. If the position of one of the objects is known, the position of the other is known as soon as their *relative* positions are determined.

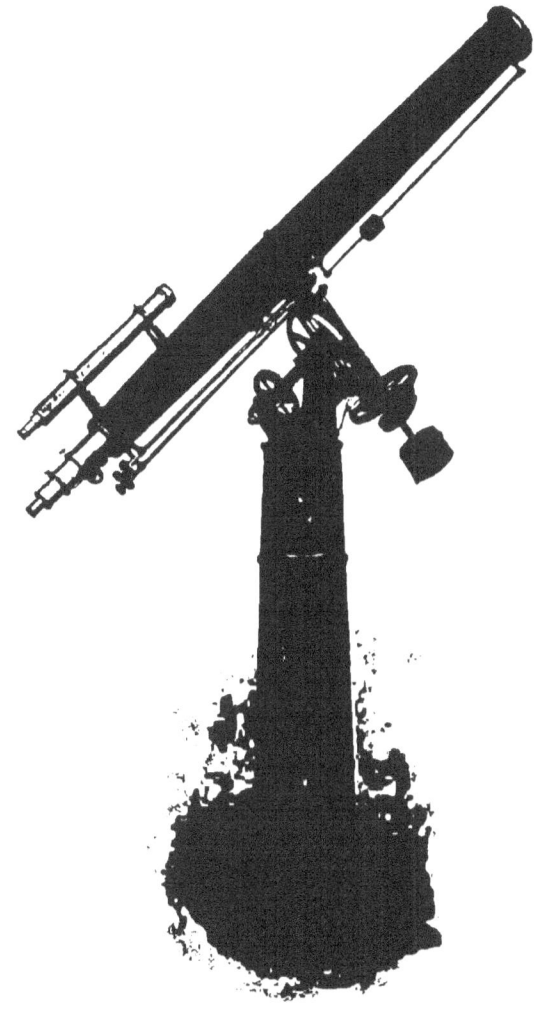

Fig. 27

**143.** By the above system of mounting it is evident that the telescope can be directed to any part of the sky; and that it will follow a star in its diurnal motion, by revolving the instrument about the polar axis alone; for in that case the line of sight maintains a constant angle with the celestial equator, and therefore describes a circle which is identical with the star's diurnal circle. Since the star's angular motion is uniform, the telescope is made to follow it by means of the sidereal driving clock. In some observations the driving clock is not used; in others it is indispensable.

When the telescope is revolved upon the declination axis, its line of sight describes an hour circle on the celestial sphere. The position of this hour circle is indicated by the reading of the graduated hour circle of the instrument. The position of the telescope in this hour circle is indicated by the reading of the graduated declination circle. When the telescope is directed exactly to the south point of the equator, the hour circle reading should be $0^h\ 0^m\ 0^s$, and the declination circle reading should be $0°\ 0'\ 0''$. Then, if the other parts of the instrument are in adjustment, and the telescope is directed to a star, the hour angle and declination of the star will be indicated (neglecting the refraction and parallax), by the hour circle and declination circle readings.*

### ADJUSTMENTS

**144.** It is not essential that the errors of adjustment of an equatorial be entirely eliminated, or that their values be accurately known; but it is a practical convenience to have the errors small, particularly so for observations on objects near the poles of the equator.

It is expected that the maker of the instrument will

---

\* The hour circle should read *time*. It should be graduated from $0^h$ to $24^h$ toward the west, or from $0^h$ to $12^h$ in both directions from $0^h$. The declination circle will read *arc*. It may be graduated from $0°$ to $360°$, or from $0°$ to $180°$ in both directions from one of its equator points, or from $0°$ to $90°$ in both directions from its two equator points. We shall suppose it to read from $0°$ to $360°$.

adjust the various parts of it as perfectly as possible *with reference to each other*. It remains for the observer to place the instrument as a whole in correct position.

The polar axis should be in the plane of the meridian; the elevation of the polar axis should equal the latitude of the place; the hour circle should read zero when the telescope is in the meridian; the declination circle should read zero when the telescope is in the equator; and the lines of sight of the finder and telescope should be parallel.

The instrument should first be placed as nearly as possible in position, by estimation. Then direct the telescope to any known star near the southern horizon whose right ascension is $\alpha$. The star will be on the meridian at the sidereal time $\theta = \alpha$. Move the whole instrument in azimuth so that the star is in the center of the telescope when the chronometer time plus the chronometer correction is equal to $\theta = \alpha$. The order of making the final adjustments is as below.

**145.** *To adjust the finder.* Using the lowest power eyepiece, direct the telescope to a bright star. Replace the low-power eyepiece by a high-power. Keeping the star in the center of the field of view, turn the adjusting screws of the finder so that the star is on the intersection of the cross wires in the finder. The two telescopes will then be sufficiently near parallelism.

**146.** *To determine the angle of elevation of the polar axis, and the index correction of the hour circle.*\* Across the object end of the telescope firmly tie a piece of wood which projects several inches from the telescope tube on the side opposite the pier. Pass a fine thread through a very small hole in the projecting end, and fasten it. Direct the telescope to the zenith. Near the eye end and on the same

---

\* This very simple and satisfactory method was proposed by Professor Schaeberle, in *der Astronomische Nachrichten*, No. 2374. It has the advantage that the errors can be determined, and corrected, in the daytime.

side as the projecting arm, fasten a block of wood. To this screw a metal plate so that it will be perpendicular to the axis of the tube, and in which is a very small circular hole as nearly as possible (by estimation) under the hole above. Pass the thread through it, tie a plumb-bob to the end of the thread near the floor, and let it swing in a vessel of water. Move the telescope by the slow-motion screws until the plumb-line passes through the center of the lower hole. Read both verniers of the hour and declination circles. Unclamp, hold the plumb-bob in the hand to avoid displacing the metal plate, reverse the telescope to the other side of the pier, and set it so that the plumb-line again passes centrally through the hole. Read both circles as before.

Let $h$ equal the angle of elevation of the polar axis. The difference of the readings of the declination circle in the two positions is $180° - 2h$. The elevation should equal the known latitude $\phi$. The error is $h - \phi$. Change the last circle reading by this amount, by moving the telescope in declination in the proper direction. Adjust the angle of elevation by the proper screws until the plumb-line again passes through the center of the hole.

If the declination circle is graduated so as to read from 0° to 90° in both directions from its two equator points, then the mean of the circle readings for the two positions of the telescope is at once the inclination of the polar axis to the horizon.

The mean of the hour circle readings in the two positions is the reading of the circle when the telescope is in the meridian. This should be $0^h\ 0^m\ 0^s$. The index error of the hour circle is the mean of the readings minus $0^h\ 0^m\ 0^s$ (or minus $24^h\ 0^m\ 0^s$). To correct for it, set the circle at this mean reading; then move the vernier screws until the reading is $0^h\ 0^m\ 0^s$.

The index correction of the hour circle is equal to the index error with its sign changed. If the error is not removed by adjusting the verniers, the index correction must be applied to every reading made with the hour circle, in order to obtain the true reading.

If the errors are large, these adjustments should be repeated once or twice.

*Example.* 1891 Feb. 20. The following plumb-line observations were made on the 6-inch equatorial of the Detroit Observatory. Determine the errors. The last column gives the position of the telescope with reference to the pier:

| Hour Circle | | Declination Circle | | Telescope |
|---|---|---|---|---|
| Vernier A | Vernier B | Vernier A | Vernier B | |
| $24^h$ $2^m$ $53^s$ | $12^h$ $2^m$ $58^s$ | $135°$ $45'$ $30''$ | $315°$ $45'$ $00''$ | E |
| 11  56  56 | 23  57  6 | 40  16  45 | 220  16  30 | W |

The means of the declination circle readings were $135°\ 45'\ 15''$ and $40°\ 16'\ 38''$, and therefore $h$ equaled $42°\ 15'\ 42''$. The value of $\phi$ is $42°\ 16'\ 47''$. The axis was therefore $1'\ 5''$ too low. The telescope was moved in declination until the verniers read $40°\ 17'\ 45''$ and $220°\ 17'\ 30''$, and the axis adjusted until the thread was again central in the hole.

The hour circle readings were $24^h\ 2^m\ 55^s.5$ and $23^h\ 57^m\ 1^s$, and their mean was $23^h\ 59^m\ 58^s.2$. The error was therefore $-1^s.8$. The verniers were moved to the west $2^s$.

A repetition of the observations gave $h = 42°\ 16'\ 49''$, and the mean of the hour circle readings, $24^h\ 0^m\ 0^s.5$. Further adjustment was not required. The index error of the hour circle was $+ 0^s.5$, and the index correction to be applied to future readings was $- 0^s.5$.

**147.** *To determine the azimuth correction of the vertical plane containing the polar axis.* This is best determined by observations on one of the four Ephemeris circumpolar stars near its culmination.

Using the micrometer eyepiece (§ 60), direct the telescope to the star a few minutes before its culmination, note the chronometer time $\theta_1$ when the star is on the point of intersection of the wires (or any well defined point in the eyepiece), and read the hour circle. Reverse the telescope to the other side of the pier, note the time $\theta_2$ when the star is at the same point of the eyepiece, and read the hour circle. Let $t_1$ and $t_2$ be the hour circle readings in the two positions, corrected for index error, if any; let $a$ be the required azimuth correction; and let $\Delta\theta$ be the known chronometer correction [see § 152].

Neglecting the quantities which are eliminated by the reversal, we have for the sidereal times when the star is in the vertical plane of the polar axis,

$$\theta_1 + \Delta\theta - t_1,$$
$$\theta_2 + \Delta\theta - t_2.$$

Therefore, as with the transit instrument, § 98,

$$aA = a - (\theta_1 + \Delta\theta - t_1),$$
$$aA = a - (\theta_2 + \Delta\theta - t_2),$$

in which $A$ is given by, (222) and (234),

$$A = \frac{\sin(\phi \mp \delta)}{\cos \delta},$$

the lower sign being for lower culmination. Solving for $a$ we find

$$a = [a - \tfrac{1}{2}(\theta_1 + \theta_2) - \Delta\theta + \tfrac{1}{2}(t_1 + t_2)]\frac{\cos \delta}{\sin(\phi \mp \delta)}. \tag{313}$$

$a$ is expressed in time: in arc, the azimuth correction is $15\,a$.

If $a$ is $+$, the south end of the axis requires to be moved to the west; if $-$, to the east. This is readily done. Direct the telescope to a distant terrestrial object nearly in the horizon, make the movable micrometer wire vertical, and set it on the object. Next move the wire through the distance $a$ in the proper direction. This can be done when the value of one revolution of the screw is known [§ 61]. Shift the whole instrument in azimuth by the proper screws until the micrometer wire is again on the object. The vertical plane of the polar axis should now coincide with the meridian.

If the value of $a$ is large the observations should be repeated. If $a$ is less than $3^s$, it will cause no inconvenience and scarcely need be corrected.

*Example.* Wednesday, 1891 February 25. 51 *Cephei* was observed at upper culmination with the 6-inch equatorial of the Detroit Observatory, as below. Determine the azimuth correction. The value of $\Delta\theta$ was $+14^m 36^s.0$.

| Telescope | Hour circle | Chronometer |
|---|---|---|
| West | $23^h 56^m 31^s$ | $6^h 31^m 42^s$ |
| East | 24  0  42 | 6  35  29 |

The index correction of the hour circle was $-0^s.5$.

| Amer. Ephem., p. 303, $\alpha$ | $6^h 49^m 27^s.5$ | $\delta$ | $+87° 13'$ |
|---|---|---|---|
| $\frac{1}{2}(\theta_1 + \theta_2)$ | 6  33  35 .5 | $\phi$ | $+42\ 17$ |
| $\Delta\theta$ | $+\ 14\ 36 .0$ | $\cos\delta$ | 8.6863 |
| $\frac{1}{2}(t_1 + t_2)$ | 23  58  36 .0 | $\sin(\phi-\delta)$ | $9.8490_n$ |
| | $-\ 8 .0$ | $\dfrac{\cos\delta}{\sin(\phi-\delta)}$ | $-0.069$ |

The value of $a$ was $-0.069 \times -8^s.0 = +0^s.6 = +9''$; that is, the south end of the axis should be moved $9''$ west. This was too small to require correction.

**148.** *To adjust the declination verniers.* Direct the telescope to a star, nearly in the zenith, whose declination is known. Bring the star to the center of the eyepiece, using a high power, and clamp the instrument in declination. Set the verniers so that they read the star's declination. They will then be in adjustment.

**149. To center the object glass.** Imperfect images are often due to the fact that the object glass is not properly centered. To test this adjustment, remove the eyepiece and hold a candle flame in such a position that the images of the flame reflected from the surfaces of the object glass can be seen through the flame. If the object glass is perfectly centered all the images should coincide when the observer's eye and the center of the flame are in the axis of the telescope. If they do not coincide, raise one side of the object glass cell by the set screws until the coincidence is perfect.

**150. The magnifying power** of a telescope is equal to the focal length of the objective divided by the focal length of the eyepiece. It is therefore different for different eyepieces on the same telescope, or for the same eyepiece on different telescopes. The following method of determining it is simple, and abundantly accurate.

Focus the telescope on a distant object, and direct it in the daytime to the bright sky. Hold a piece of thin, unglazed paper in front of the eyepiece at such a distance that the bright disk formed on it is clearly defined. This disk is the minified image of the object glass. Measure its diameter by a finely divided scale held against the paper, and measure the diameter of the object glass. It can be shown that these diameters are to each other as the focal lengths of the eyepiece and object glass. Their quotient is therefore the magnifying power. Thus, for the equatorial mentioned above, the diameter of the object glass is 6.05 inches, and the diameter of the bright disk for a certain eyepiece is 0.08 inch. The magnifying power is, therefore, $6.05 \div 0.08 = 76$.

A definite statement of the magnifying power to be used in observing an object cannot be made. A higher power can be used when the *seeing* is good, *i.e.*, when the images in the telescope are steady and well defined, than when

CHRONOMETER CORRECTION 221

the seeing is poor. Lower powers must in general be used with nebulæ and comets. The very highest powers can be used with stars and some of the planetary nebulæ, if the seeing is good. Further than this, the observer must select that eyepiece which on trial gives the best results.

**151. The field of view** is the circular portion of the sky which can be seen through the telescope at one time. Its diameter is equal to the angle contained by two rays drawn from the center of the object glass to the two extremities of a diameter of the eyepiece. The diameter, expressed in arc, is equal to 15 times the interval of time required by an equatorial star to traverse it. This can be directly observed.

**152.** *The chronometer correction* is quickly obtained, with an accuracy sufficient for all ordinary uses of the equatorial, by the following method:

Direct the telescope to a known star nearly in the zenith, note the chronometer time $\theta_1$ when the star is on the point of intersection of the wires, and read the hour circle. Carry the telescope to the other side of the pier, observe the star as before at the time $\theta_2$, and read the hour circle. Let the hour circle readings corrected for index error be $t_1$ and $t_2$. We have, by (40),

$$a = \theta_1 + \Delta\theta - t_1,$$
$$a = \theta_2 + \Delta\theta - t_2,$$

neglecting only very small quantities and those eliminated by reversal. Therefore

$$\Delta\theta = a - \tfrac{1}{2}(\theta_1 + \theta_2) + \tfrac{1}{2}(t_1 + t_2). \tag{314}$$

For many purposes an observation on one side of the pier will suffice, and we have

$$\Delta\theta = a + t_1 - \theta_1. \tag{315}$$

*Example.* Ann Arbor, Wednesday, 1891 Feb. 25. The following observation of *Castor* was made with the 6-inch equatorial, to determine an approximate value of the chronometer correction.

Chronometer time,    $\theta_1$   7   $5^m$   $9^s$
Hour circle,           $t_1$ 23   52   3
Amer. Ephem., p. 327,   $a$   7   27   39

Therefore, by (315), $\Delta\theta = + 14^m 33^s$.

**153.** *To direct the telescope to an object* whose right ascension ($a$) and declination ($\delta$) are known, first determine whether the object is east or west of the meridian. If the right ascension is greater than the sidereal time, it is east; if less, it is west. Generally, if the object is east, the telescope should be west of the pier; if the object is west, the telescope should be east of the pier. Move the telescope in declination till the declination circle reads $\delta$. To the reading of the chronometer add the chronometer correction, and one or two minutes more for the time consumed in setting. Subtract $a$ from this sum and set off the difference (which is the hour angle) on the hour circle. When the chronometer indicates the time for which the hour angle was computed, the object should be seen in the finder. Move the telescope until the star is on the intersection of the finder cross-wires. The star should then be visible near the center of the field of view of the (principal) telescope.

Conversely, if an unknown star is seen in the telescope, the chronometer time noted and the circle readings taken: then the declination circle reading is the star's declination; and the chronometer time of observation, plus the chronometer correction, minus the hour circle reading, is its right ascension.

These results are only approximate, of course, since the instrument will never be in perfect adjustment, and the star will not be seen in its true place, owing to the refraction, etc.

## DETERMINATION OF APPARENT PLACE OF AN OBJECT

**154.** *By the method of micrometer transits.* Select a known * star, called a **comparison star**, whose right ascension and especially whose declination differ as little as possible from that of the object. Revolve the filar micrometer [§ 60] until the star in its diurnal motion follows along the micrometer wire. The wire in this position is exactly east and west, or parallel to the equator, and the reading of the position circle for this direction of the wires is called the **equator reading**. If the object and comparison star are in the vicinity of the pole, their diurnal circles will be sharply curved, and in this case the equator reading of the circle should, first of all, be determined from an equatorial star. Direct the telescope just in advance of the two objects. The diurnal motion will carry them across the field. Note the chronometer times when they cross the transverse wire or wires. The difference of these times for the two objects is the difference of their right ascensions. Also, when the first or *preceding* object approaches the center of the field, move the whole system of wires until the object follows along the fixed wire. When the second or *following* object approaches the center of the field, bisect it with the micrometer wire. Read the micrometer in this position, and also when the micrometer wire is in coincidence with the fixed wire. The difference of the two readings, multiplied by the value of a revolution of the screw [§ 61], is the difference of the declinations of the two objects.

Care should be taken to have the micrometer and fixed wires exactly parallel,† and the transverse wire (or wires) exactly perpendicular to the micrometer wire. To test the

---

\* That is, a star whose accurate position is given in one or more of the star catalogues.

† In good forms of the micrometer, an adjusting screw is provided for bringing them into parallelism.

relative positions of the two sets of wires, direct the telescope to an equatorial star, adjust the micrometer wire so that the star's diurnal motion will carry it across the field in coincidence with the wire, and read both verniers of the position circle. Rotate the system of wires, adjust the transverse wire so that the star will cross the field in coincidence with the wire, and read both verniers. The difference of the two position circle readings should be exactly 90°.

Many observers prefer to observe only one coördinate at a time. A good program to follow is, measure the difference of declination, revolve the micrometer 90° and observe the difference of right ascension, then revolve 90° more and measure the difference of declination again.

In any case, the observations should be repeated several times, and the mean of all the observations adopted. If the object has a proper motion, the differences in right ascension and declination are those corresponding to the *instant when that object was observed:* that is, the mean of the chronometer times for the object, plus the chronometer correction.

The mean place of the comparison star * will be given for the epoch of the catalogue which contains it. Reducing this to the mean place for the beginning of the year of observation by § 46, 47 or 52, thence to the apparent place for the instant of observation by § 55, and applying the micrometer differences to the apparent place, we obtain the observed place of the object. This must be corrected for refraction and parallax.

The refraction correction will be small, since the star

---

\* If the star is a very bright one, it may be identified satisfactorily, both in the sky and in the catalogue, by the methods of § 153. But, in case it is faint, the observer should always compare a chart of the neighboring stars, prepared from the catalogue, with that region of the sky, making sure that the configurations of the stars on the chart and in the sky agree.

and object will be refracted nearly the same amount in nearly the same direction. An equatorial is a fixed part of an observatory, and tables of differential refractions in right ascension and declination in every part of the sky should be computed for the latitude of the observatory. The corrections can then be taken from the tables very quickly.

Until such tables are constructed, the corrections can be computed by the following method: Let $t_0$ be the mean of the hour angles of the star and object, $\delta_0$ the mean of their declinations, and $z$ the mean of their zenith distances. Substitute these values of $t$ and $\delta$ in (35), (36) and (37), to determine $L$ and $z$. The corrections to the observed place will be given by *

$$\Delta\alpha = \frac{\kappa}{15} \cdot \frac{\delta' - \delta}{\sin^2(\delta_0 + L)} \cdot \frac{\tan t_0 \sin L \cos(2\delta_0 + L)}{\cos^2 \delta_0}, \quad (316)$$

$$\Delta\delta = \kappa \cdot \frac{\delta' - \delta}{\sin^2(\delta_0 + L)}, \quad (317)$$

in which $\delta' - \delta$ is the declination of the object minus the declination of the star expressed in seconds of arc, and $\kappa$ is defined by

$$\kappa = \mu'' B T \gamma^{\lambda''}. \quad (318)$$

$B$, $T$ and $\gamma$ have the same significance as in § 30, and their values are given in TABLE I of the Appendix. The values of $\log \mu''$ and $\lambda''$ are tabulated below with the argument $z$.

| $z$ | $\log \mu''$ | $\lambda''$ | $z$ | $\log \mu''$ | $\lambda''$ |
|---|---|---|---|---|---|
| 0° | 6.446 | 1.00 | 80° | 6.395 | 1.10 |
| 45 | 6.444 | 1.00 | 82 | 6.370 | 1.15 |
| 60 | 6.440 | 1.01 | 83 | 6.351 | 1.18 |
| 70 | 6.433 | 1.03 | 84 | 6.323 | 1.21 |
| 75 | 6.422 | 1.05 | 85 | 6.285 | 1.24 |

* These equations are derived in *Chauvenet's Spherical and Practical Astronomy*, Vol. II, pp. 450–460.

It is only in the most refined measurements and in extreme states of the weather that the barometer and thermometer readings need be taken into account. With comets it will scarcely ever be necessary, except when they are very near the horizon.

If one or both of the bodies is in the solar system, and at different distances from the observer, the observations will require correction for parallax, by the methods explained in full in § 28.

Four-place tables are sufficient for computing the refraction and parallax corrections.

*Example.* Wednesday, 1890 July 23, the author made the following observation of Coggia's comet with the 12-inch equatorial of the Lick Observatory. Required its apparent place.

The comet was south of and preceding the 6th magnitude star No. 1518 *Pulkowa Catalogue* for 1855.0. The reading of the position circle when the star followed along the micrometer wire was 201°.35. The micrometer readings when the micrometer and fixed wires coincided were

19.947
.947
.947
Mean 19.947

When the comet was in the center of the field the fixed wire was made to bisect it, and the chronometer time was noted. When the star reached the center of the field (nearly four minutes later) the micrometer wire was made to bisect it, and the micrometer reading noted. In this way the difference of the declinations was observed, as below.

| | Chronometer | Micrometer | Remarks |
|---|---|---|---|
| | $16^h\ 43^m\ 11^s$ | 22.954 | Very windy |
| | 48  59 | 23.822 | Very windy |
| | 54   2 | 24.550 | Very windy |
| Means | 16  48  44 | 23.775 | |

DETERMINATION OF APPARENT PLACE 227

The micrometer was rotated 90° until the circle read 291°.35, and the chronometer times of transit over the two wires noted, as below.

| Comet | | | Star | | Difference |
|---|---|---|---|---|---|
| $17^h$ $0^m$ 51$^s$.7 | $0^m$ 59$^s$.2 | $17^h$ $4^m$ 44$^s$.6 | $4^m$ 51$^s$.8 | | $-3^m$ 52$^s$.75 |
| 5 42.3 | 5 49.5 | 9 34.0 | 9 41.3 | | 3 51.75 |
| 10 37.8* | 10 46.1 | 14 28.0 | 14 36.3 | | 3 50.20 |
| 23 3.5 | 23 11.5 | 26 49.8 | 26 58.1 | | 3 46.45 |
| 27 39.5* | 27 52.2 | 31 24.6 | 31 37.2 | | $-3$ 45.05 |
| Means $17^h$ $13^m$ $39^s$ | | | | | $-3$ 49.24 |

The micrometer was rotated 90° further until it read 21°.35, and the difference of the declinations measured again, as below,

| Chronometer | Micrometer | Remarks |
|---|---|---|
| $17^h$ $34^m$ $22^s$ | 9.751 | Very windy |
| 39 44 | 8.867 | Very windy |
| 48 40 | 7.545 | Very windy |
| Means 17 40 55 | 8.721 | |

Readings for coincidence of wires,

19.944
.946
.943
Mean 19.944

The value of one revolution of the screw is $R = 14''.058$. We shall combine the two differences of declination, thus:

| Chronometer | Diff. of decl. |
|---|---|
| $16^h$ $48^m$ $44^s$ | $- 3.828 R$ |
| 17 40 55 | $- 11.223 R$ |
| Means 17 14 49 | $- 7.525 R = - 1' 45''.8$. |

The chronometer time for the declinations is $1^m 10^s$ greater than that for the right ascensions. From the two measured declinations it is found that the declination changed $2''.3$ in $1^m 10^s$. Therefore, at $17^h 13^m 39^s$ the difference of declination was $- 1' 43''.5$.

---

* The distance between the wires was changed intentionally.

The mean place and proper motion of the comparison star for 1855.0 given by the catalogue were

$$\alpha = 9^h\,29^m\,17^s.57, \quad \delta = +\,40°\,53'\,16''.1,$$
$$\mu = -\,0^s.0022, \quad \mu' = +\,0''.008.$$

The mean place for 1890.0 was, by § 47,

$$\alpha = 9^h\,31^m\,29^s.56, \quad \delta = +\,40°\,43'\,58''.8\,;$$

and the apparent place for sidereal time 1890 July $23^d\,17^h$, by § 55,

$$\alpha' = 9^h\,31^m\,28^s.90, \quad \delta' = +\,40°\,44'\,4''.2.$$

Therefore the observed place of the comet at $17^h\,13^m\,39^s$ was

$$\alpha = 9^h\,27^m\,39^s.66, \quad \delta = +\,40°\,42'\,20''.7.$$

The chronometer correction was $-\,1^m\,19^s$. We have

| | | | | |
|---|---|---|---|---|
| Chronometer time, | $\theta'$ | $17^h$ | $13^m$ | $39^s$ |
| Chronometer corr., | $\Delta\theta$ | — | 1 | 19 |
| Sidereal time, | $\theta$ | 17 | 12 | 20 |
| Right ascension, | $\alpha$ | 9 | 27 | 40 |
| Hour angle, | $t$ | 7 | 44 | 40 |

The corrections for differential refraction corresponding to this hour angle and declination, taken from the tables constructed for the Lick Observatory, or computed from (316) and (317), are

$$\Delta\alpha = -\,0^s.04, \quad \Delta\delta = -\,0''.6.$$

The corrections for parallax at the unit distance, *i.e.* the parallax factors, taken from tables constructed for the Lick Observatory, or computed from (92), (90) and (93), are

$$\Delta\alpha = +\,0^s.56, \quad \Delta\delta = +\,6'\,.1.$$

The comet's distance was 1.57, and therefore the required corrections for parallax are

$$\Delta\alpha = +\,0^s.36, \quad \Delta\delta = +\,3\,.9.$$

DETERMINATION OF APPARENT PLACE    229

Applying these corrections to the observed place we obtain the following apparent place of the comet:

Mt. Hamilton sid. time   Apparent $\alpha$        Apparent $\delta$
1890 July 23, $17^h\ 13^m\ 39^s$   $9^h\ 27^m\ 39^s.98$   $+ 40°\ 42'\ 24''.0$

**155.** *By the method of direct micrometer measurement.* When the object whose position, $(\alpha'', \delta'')$, is to be determined is comparatively near the comparison star, $(\alpha', \delta')$, so that both are well within the field of view, their differences of right ascension and declination are conveniently determined by direct micrometer measurement.

Let the micrometer wire be placed parallel to the equator, as before. By means of the driving clock keep the telescope directed to the star and object so that the point midway between them will be *as nearly as possible in the center of the field.* Bisect the star's image with the fixed wire and the object's image with the micrometer wire, and note the chronometer time and the reading $m$ of the micrometer. If $m_0$ is the reading for coincidence of the wires, and $R$ the value of a revolution of the screw, then the apparent difference $\Delta\delta$ of the declinations is given by

$$\Delta\delta = (m - m_0)R. \qquad (319)$$

Rotate the wires through 90°, bisect the two images as before and note the time and the micrometer reading $m'$. The apparent difference of the right ascensions is given by

$$\Delta\alpha = (m' - m_0) R \sec \tfrac{1}{2}(\delta'' + \delta'). \qquad (320)$$

This method cannot be used with safety near the pole unless the instrument is in good adjustment and the difference of right ascension is small.

The apparent differences of right ascension and declination will require correction for differential refraction. The corrections could be computed by differential formulæ, but an equally satisfactory method consists in computing the absolute refractions in right ascension and declination for the star and object, and taking their differences as the corrections to the observed intervals $\Delta a$ and $\Delta \delta$. The values of the parallactic angles and zenith distances, $q'$, $z'$ for the star and $q''$, $z''$ for the object, may be taken from general tables constructed for the point of observation, or computed by (35), (36) and (37). These should be partially checked by the formula

$$z'' - z' = \Delta z = -\cos \delta' \sin q' \Delta a - \cos q' \Delta \delta, \qquad (321)$$

formed by differentiating (30) and (41) and combining the results with (31). The refraction $r'$ for the star may be computed by means of (95), (96) or (97), remembering that $z$ in these formulæ is the apparent zenith distance. Formula (96) will be sufficiently accurate, except in case of observations made very near the horizon. The refraction $r''$ for the object should now be found differentially. The change $\Delta r$ in refraction due to a change $\Delta z$ in zenith distance is obtained from (96), thus:

$$\Delta r = \frac{983\,b}{460 + t} \sec^2 z \sin 1'' \Delta z, \qquad (322)$$

in which $\Delta r$ and $\Delta z$ are expressed in terms of the same unit. The refraction for the object will be given by

$$r'' = r' + \Delta r. \qquad (323)$$

The corrections to the apparent places may be obtained from (100) and (101), thus:

$$da' = -r' \sin q' \sec \delta', \qquad d\delta' = -r' \cos q', \qquad (324)$$
$$da'' = -r'' \sin q'' \sec \delta'', \qquad d\delta'' = -r'' \cos q''. \qquad (325)$$

Therefore, we shall have the true differences of right ascension and declination, as seen from the point of observation,

$$a'' - a' = \Delta a + (da'' - da'), \qquad (326)$$
$$\delta'' - \delta' = \Delta \delta + (d\delta'' - d\delta'), \qquad (327)$$

from which the values of $a''$ and $\delta''$ may be obtained.

If the object is in the solar system, it will further require correction for parallax.

DETERMINATION OF APPARENT PLACE     231

*Example.* Lick Observatory, Friday, 1898 Nov. 11. 12-inch equatorial. Observer, W. J. Hussey. The differences of right ascension and declination of the minor planet *Eros* and the star DM. $- 4°.5413$ were measured directly with the micrometer, to determine the apparent place of the planet.

The mean of five measures of difference of declination was, (the planet being north of the star),

$m = 55^r.007$ at (sidereal) chronometer time $22^h\ 48^m\ 21^s.0$.

The mean of ten measures of difference of right ascension was, (the planet being west of the star),

$m' = 67^r.180$ at chronometer time $22^h\ 52^m\ 22^s.7$.

The mean of five additional measures of difference of declination was

$m = 55^r.228$ at chronometer time $22^h\ 56^m\ 27^s.2$.

The coincidence of the wires was at $46^r.650$; the value of one revolution of the screw is $14''.058$; the chronometer correction was $+ 2^m 53^s.7$; and the apparent place of the star for the instant of observation was [from *Karlsruhe Beobachtungen*],

$a' = 21^h 13^m 10^s.03, \qquad \delta' = - 4° 6' 10''.9$.

The observations for determining $\Delta \delta$ will be combined as follows, the reductions $- 1^s.4$ and $- 0^r.001$ being applied so that the declination observations will refer to the same instant as the right ascension observations.

| | | | |
|---|---|---|---|
| Chronometer, | $22^h\ 48^m\ 21^s.0$ | $m,$ | $55^r.007$ |
| | $22\ \ 56\ \ 27.2$ | | $55.228$ |
| Means, | $22\ \ 52\ \ 24.1$ | | $55.118$ |
| Reductions, | $- 1.4$ | | $- .001$ |
| | $22\ \ 52\ \ 22.7$ | | $55.117$ |
| | | $M_0,$ | $46.050$ |
| | | $\Delta \delta,$ | $+ 8.467\ R = + 119''.04$ |
| Chronometer, | $22\ \ 52\ \ 22.7$ | $m',$ | $67.180$ |
| Chron. corr., | $+ 2\ \ 53.7$ | $m_0,$ | $46.050$ |
| Sid. time, $\theta,$ | $22\ \ 55\ \ 16.4$ | $\Delta a,$ | $- 20.530\ R \sec(- 4° 5.2)$ |
| | | $\Delta a,$ | $- 289''.36$ |

Applying these values of $\Delta\alpha$ and $\Delta\delta$ to the star's place we have the approximate position of the planet,

$$\alpha'' = 21^h\, 12^m\, 51^s.3, \qquad \delta'' = -4°\, 4'\, 12''.$$

The data for solving (35), (36) and (37) are

$$\begin{aligned}
\phi &= \phantom{+}37°\, 20'\, 26'' & \theta &= \phantom{+}22^h\, 55^m\, 16^s.4 \\
t' &= +\phantom{0}1^h 42^m\, 6^s & t'' &= +\phantom{0}1\phantom{°}\, 42\phantom{'}\, 25 \\
l' &= +25°\, 31'\, 30'' & l'' &= +25°\, 36'\, 15'' \\
\delta' &= -\phantom{0}4\phantom{°}\, \phantom{0}6\phantom{'}\, 11 & \delta'' &= -\phantom{0}4\phantom{°}\, \phantom{0}4\phantom{'}\, 12
\end{aligned}$$

The quantities obtained from the solution are

$$\begin{aligned}
q' &= 27°\, 33'\, 50'' & q'' &= 27°\, 38'\, 46'' \\
z' &= 47\phantom{°}\, 45\phantom{'}\, 42 & z'' &= 47\phantom{°}\, 46\phantom{'}\, 10
\end{aligned}$$

The value of $z'' - z' = \Delta z = +28''$ agrees exactly with that obtained from (321).

The mean value of the refraction at true zenith distance $z' = 47°\, 46'$ is about $1'$. Solving (96), (322) and (323) for $z = 47°\, 45'$ and $\Delta z = +28''$, assuming $b = 25.8$ inches and $t = +55°$ as the average for observing weather at that season of the year, we obtain

$$\begin{aligned}
r' &= 54''.2, \\
r'' &= 54\phantom{'}.2 + 0''.015 = 54''.215.
\end{aligned}$$

Substituting these in (324) and (325) we obtain

$$\begin{aligned}
d\alpha' &= -25''.14, & d\delta' &= -48''.05, \\
d\alpha'' &= -25\phantom{''}.22, & d\delta'' &= -48\phantom{''}.03;
\end{aligned}$$

and, therefore, from (326) and (327),

$$\begin{aligned}
\alpha'' - \alpha' &= -289''.36 - 0''.08 = -289''.44 = -19^s.30, \\
\delta'' - \delta' &= +119\phantom{''}.04 + 0\phantom{''}.02 = +119\phantom{''}.06 = +\phantom{0}1'\, 59''.1;
\end{aligned}$$

whence

$$\alpha'' = 21^h\, 12^m\, 50^s.33, \qquad \delta = -4°\, 4'\, 11''.8.$$

The distance $\Delta$ of the planet from the earth being 1.225 units, the parallax corrections taken from general tables or computed from (89), (90) and (91), are $+0^s.17$ and

$+ 4''.7$. The apparent place of the planet referred to the center of the earth was therefore

| Mt. Hamilton sidereal time | Apparent $\alpha$ | Apparent $\delta$ |
| --- | --- | --- |
| 1898 Nov. 11, $22^h\ 55^m\ 16^s.4$ | $21^h\ 12^m\ 50^s.50$ | $-4°\ 4'\ 7''.1$ |

DETERMINATION OF POSITION ANGLE AND DISTANCE

**156.** The relative positions of two objects close together are conveniently expressed in terms of position angle and distance. The **position angle**, $p$, of one star, $B$, with reference to another star, $A$, is the angle which the great circle passing through the two objects makes with the hour circle passing through $A$, reckoned from the north toward the east through 360°. Their **distance**, $s$, is the length of the arc of the great circle joining them.

To determine the position angle, revolve the micrometer until one of the stars, by its diurnal motion, follows along the micrometer wire, and note the reading $P_0$ of the position circle. Keeping the telescope directed upon the stars by means of the driving clock, revolve the micrometer until the micrometer wire passes through the two stars, and note the circle reading $P$. The position angle required is

$$p = P - (P_0 \pm 90°). \tag{328}$$

To determine the distance, revolve the micrometer to the circle reading $P \pm 90°$. Bisect one of the stars with the fixed wire by moving the whole system of wires, then bisect the other star with the micrometer wire, and note the reading $m$ of the micrometer. If $m_0$ is the reading of the micrometer when the two wires coincide, and $R$ the value of a revolution of the screw, the required distance is

$$s = (m - m_0) R. \tag{329}$$

In very accurate measurements bisect the two stars as above, and take the reading $m$. Move the micrometer wire to the other side of the fixed wire, bisect the stars with the wires in that order, and take the reading $m'$. The distance is now given by

$$s = \tfrac{1}{2}(m' - m)R. \tag{330}$$

In this way several systematic and personal errors are eliminated, partially at least. This method is called **the method of double distances**.

The values of $s$ and $p$ will be affected by refraction; but in the case of double stars, to which the method is especially applied, the correction for refraction may usually be neglected.

If the distance between the two stars is large, the telescope should be directed so that the two stars will fall on opposite sides of the center of the field, and at equal distances from the center. In this case the measured position angle is the angle between the arc joining the two stars, and the hour circle passing through the middle point of that arc. Let $p'$ and $s'$ be the observed angle and distance. Let their values corrected for refraction be $p$ and $s$. Let $z$ and $q$ be determined for the point midway between the two stars from (35), (36) and (37), and let $\kappa$ be defined as in (318). It can be shown * that

$$p = p' - \kappa \csc 1'' [\tan^2 z \cos(p' - q) \sin(p' - q) \\ - \tan z \sin q \tan \tfrac{1}{2}(\delta + \delta')], \tag{331}$$

$$s = s' + s\kappa [\tan^2 z \cos^2(p' - q) + 1]. \tag{332}$$

This value of $p$, referring to the point midway between the two stars, may readily be converted into the position angle of each star with reference to the other star. Let $S'$, in the position $a'$, $\delta'$, represent the western star; $S''$, in the position $a''$, $\delta''$, the eastern star; $M$ the point midway between them; and $P$ the pole of the equator. Let $p'$ be the position angle of the eastern star with reference to the western, and $180° + p''$ the position angle of the

---

* See *Chauvenet's Sph. and Prac. Astronomy*, Vol. II, pp. 450–459.

western star with reference to the eastern. In practice, the declination of one star will always be known; and if the declination of the other is unknown, its value may be found with sufficient accuracy from equation (338) below. Let $\delta_0 = \frac{1}{2}(\delta'' + \delta')$. Without sensible error, we may assume $\delta_0$ as the declination of the point $M$ defined above. Then, from the triangles $PS'M$ and $PS''M$ we can write [*Chauvenet's Sph. Trig.*, § 70, (N)],

$$\tan p' = \frac{\sin p}{\cos \frac{1}{2} s \cos p + \sin \frac{1}{2} s \tan \delta_0}, \tag{333}$$

$$\tan p'' = \frac{\sin p}{\cos \frac{1}{2} s \cos p - \sin \frac{1}{2} s \tan \delta_0}, \tag{334}$$

which determine $p'$ and $p''$. Their values will differ very little from $p$, unless the stars are very near the pole.

It is frequently required to convert position angle and distance into the corresponding differences of right ascension and declination. From the triangle $PS'S''$ defined above, we may write [*Chauvenet's Sph. Trig.*, (44)],

$$\sin \tfrac{1}{2}(a'' - a') = \sin \tfrac{1}{2} s \sin \tfrac{1}{2}(p'' + p') \sec \delta_0, \tag{335}$$

$$\sin \tfrac{1}{2}(\delta'' - \delta') = \sin \tfrac{1}{2} s \cos \tfrac{1}{2}(p'' + p') \sec \tfrac{1}{2}(a'' - a'), \tag{336}$$

which solve the problem. If the stars are at some distance from the pole, we may safely substitute $p$ for $\frac{1}{2}(p'' + p')$ in these equations.

If the stars are not far from the equator, or if $s$ is relatively small, or if only moderate accuracy is required, these equations may be written,

$$a'' - a' = s \sin p \sec \delta_0, \tag{337}$$

$$\delta'' - \delta' = s \cos p. \tag{338}$$

*Example.* 36-inch equatorial, Lick Observatory, Thursday night, 1898 November 17. Observer, W. J. Hussey. The position angle and distance of the faint companion of *Sirius* with reference to the principal component were measured as below. The distance was determined by the method of double distances. The equator reading of the position circle was 111°.7, and the value of a revolution of the micrometer screw is 9″.907.

| | | | | |
|---|---|---|---|---|
| $P$, | 185°.6 | $m'$, 56ʳ .199 | | $m$, 55ʳ.344 |
| | 184 .7 | .182 | | .350 |
| | 184 .3 | .180 | | .336 |
| | 184 .7 | .178 | | .342 |
| | 182 .2 | .193 | | .336 |
| | 185 .0 | Mean $m'$, 56 .186 | | Mean $m$, 55 .342 |
| | 184 .3 | $m$, 55 .342 | | |
| Mean $P$, | 184 .4 | $\frac{1}{2}(m' - m)$, 0ʳ .422 | | |
| $P_0$, | 111 .7 | $R$, 9″.907 | | |
| | 72 .7 | $s$, 4″.18 | | |
| | + 90 .0 | | | |
| $p$, | 162°.7 | | | |

The correction for refraction is not appreciable.

## THE RING MICROMETER

**157.** This consists of a narrow metal ring, one or both of its edges turned exactly circular, attached to a thin piece of glass in the focal plane of an eyepiece. When the eyepiece is put on the telescope and focused, the ring is also in the focal plane of the object glass.

If the times of transit of two stars over the edges of the ring are observed, the differences of their right ascensions and declinations can be found. But results obtained in this way can be regarded as only approximately correct, and the ring micrometer should never be used with an equatorial telescope unless, in case of great haste, there is not time to attach the filar micrometer and adjust its wires by the diurnal motion. The principal advantage of the ring micrometer is that it can be used with an instrument mounted in altitude and azimuth as well as with an equatorial, whereas a filar micrometer cannot.

**158.** *To find the radius of the ring.* Select two stars whose declinations are accurately known, the difference of whose declinations is a little less than the diameter of the ring, and whose right ascensions do not differ more than three or four minutes. Two stars to fulfil these conditions can always be found in the Pleiades. When these stars are nearly on the observer's meridian, observe their transits over the edge of the ring.

In Fig. 28 let $CDD'C'$ represent the ring; $CD$ the path of one star $(\alpha, \delta)$, and $t_1$ and $t_2$ the observed sidereal times of its transit over $C$ and $D$; $C'D'$ the path of the other star $(\alpha', \delta')$, and $t_1'$ and $t_2'$ the times of its transit over $C'$ and $D'$. Draw $MM'$ perpendicular to $CD$ and $C'D'$. Draw the radii $CO$ and $C'O$, and let $r$ represent their value in seconds of arc. If we put

$$COM = \gamma, \quad C'OM' = \gamma',$$

we can write

$$OM = r\cos\gamma, \quad CM = r\sin\gamma,$$
$$OM' = r\cos\gamma', \quad C'M' = r\sin\gamma';$$

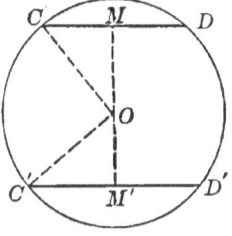

Fig. 28

and therefore

$$MM' = r(\cos\gamma' + \cos\gamma) = 2r\cos\tfrac{1}{2}(\gamma'+\gamma)\cos\tfrac{1}{2}(\gamma'-\gamma), \quad (339)$$

$$C'M' + CM = r(\sin\gamma' + \sin\gamma) = 2r\sin\tfrac{1}{2}(\gamma'+\gamma)\cos\tfrac{1}{2}(\gamma'-\gamma), \quad (340)$$

$$C'M' - CM = r(\sin\gamma' - \sin\gamma) = 2r\cos\tfrac{1}{2}(\gamma'+\gamma)\sin\tfrac{1}{2}(\gamma'-\gamma). \quad (341)$$

We have

$$MM' = \delta' - \delta, \quad CM = \tfrac{15}{2}(t_2 - t_1)\cos\delta, \quad C'M' = \tfrac{15}{2}(t_2' - t_1')\cos\delta';$$

and if we put

$$\tfrac{1}{2}(\gamma' + \gamma) = A, \quad \tfrac{1}{2}(\gamma' - \gamma) = B,$$

we can write

$$\tan A = \frac{C'M' + CM}{MM'} = \frac{\tfrac{15}{2}(t_2' - t_1')\cos\delta' + \tfrac{15}{2}(t_2 - t_1)\cos\delta}{\delta' - \delta}, \quad (342)$$

$$\tan B = \frac{C'M' - CM}{MM'} = \frac{\tfrac{15}{2}(t_2' - t_1')\cos\delta' - \tfrac{15}{2}(t_2 - t_1)\cos\delta}{\delta' - \delta}, \quad (343)$$

$$r = \frac{MM'}{2\cos A \cos B} = \frac{\delta' - \delta}{2\cos A \cos B}. \quad (344)$$

The apparent distance between the stars is affected by refraction. Since the observations are made near the meridian, the refraction in right ascension can be neglected, and it will be sufficiently accurate to consider that its effect upon the difference of the declinations is equal to the difference of refraction in zenith distance of the stars when they are on the meridian. The difference of the declinations furnished by the star catalogues requires to be decreased numerically by the difference of the refractions before substituting in the above equations. The difference of the refractions may be readily obtained with the assistance of the table in § 111, based upon equation (280), or from the table of mean refractions, Appendix, TABLE II.

*Example.* Friday, 1889 Jan. 25. The following times of transit of 23 *Tauri* and 27 *Tauri* over the *outer* edge of the ring micrometer of the 12¼-inch equatorial of the Detroit Observatory were noted, to determine the radius of the ring.

| 23 *Tauri* | 27 *Tauri* |
|---|---|
| $t_1 = 3^h\ 30^m\ 41^s.5,$ | $t_1' = 3^h\ 33^m\ 33^s.5,$ |
| $t_2 = 3\ \ 30\ \ 53\ .0,$ | $t_2' = 3\ \ 33\ \ 41\ .0.$ |

The mean places of these stars for 1850.0 are given in *Newcomb's Standard Stars.* Reducing them to the mean place for 1889.0 by § 52, and thence to the apparent place at the instant of observation by § 55, (*b*), we obtain

$$\delta = 23° 36' 4''.13, \qquad \delta' = 23° 42' 45''.40.$$

The zenith distance of the stars is about 18° and the difference of their zenith distances is about 7'. Entering the table in § 111 with $\frac{1}{2}(z' - z'') = 3'.5$ and $z = 18°$, onehalf the difference of the refractions is $0''.07$, or the whole difference is $0''.14$. Therefore the apparent difference of declination of the stars is

$$\delta' - \delta = 401''.13.$$

From (342) and (343) we find

$$A = 18° 1' 34'', \qquad B = 3° 55' 37'';$$

and then, from (344)

$$r = 211''.4.$$

The mean value of $r$ from nine complete sets of transits is

$$r = 210''.7 \pm 0''.18.$$

**159.** *To determine the difference of the right ascensions and declinations of two stars.* Observe the transits over the edge of the ring, as in § 158. Using the notation of § 158, the difference of the right ascensions is given by

$$a' - a = \tfrac{1}{2}(t_1' + t_2') - \tfrac{1}{2}(t_1 + t_2). \tag{345}$$

Letting $OM = d$, and $OM' = d'$ we can write

$$\sin \gamma = \frac{\tfrac{1}{2}^s \cos \delta}{r}(t_2 - t_1), \quad \sin \gamma' = \frac{\tfrac{1}{2}^s \cos \delta'}{r}(t_2' - t_1'), \tag{346}$$

$$d = r \cos \gamma, \qquad d' = r \cos \gamma'. \tag{347}$$

The difference of the declinations is given by

$$\delta' - \delta = d' \pm d. \tag{348}$$

The lower sign is used if the stars are observed on the same side of the center of the ring. Equations (347) do not determine the signs of $d$ and $d'$, but there will never be any ambiguity if the observer notes the positions of the two stars with reference to the center of the ring.

Differences obtained in this way are slightly in error on account of refraction; on account of the fact that the paths of the stars are arcs and not chords of the circle (except for equatorial stars); and on account of the proper motion of one of the bodies (if it have a proper motion). But as stated above, the ring micrometer should not be used on an equatorial telescope when exact measurements are required; so that the corrections for these errors will seldom be justified, and we shall not consider them here.

240   PRACTICAL ASTRONOMY

*Example.* Saturday, 1888 Sept. 8. The position of Comet *c* 1888 was compared with that of the star $13^h$, 1197 in *Weisse's Bessel's Catalogue*, by observing the transits of the star and comet over the *inner* edge of the ring micrometer of the $12\frac{1}{4}$-inch equatorial of the Detroit Observatory. The times of transit were

| Star | Comet |
|---|---|
| $t_1 = 19^h\ 18^m\ 34^s.1$, | $t_1' = 19^h\ 19^m\ 12^s.6$, |
| $t_2 = 19\ \ 18\ \ 50.0$, | $t_2' = 19\ \ 19\ \ 31.1$. |

The image of the star was north of the center and that of the comet was south. The radius of the ring is $171''.6$.
The apparent place of the star was

$$\alpha = 13^h\ 56^m\ 3^s.06, \qquad \delta = +31° 13' 49''.6.$$

The declination of the comet was approximately $+ 31° 18'$.
Substituting in (345) we find

$$\alpha' - \alpha = + 0^m\ 39^s.80.$$

The solution of (346) and (347) is here given.

| | |
|---|---|
| $\log \frac{1}{2}$  0.87506 | $\log \frac{1}{2}$  0.87506 |
| $\cos \delta$  9.93201 | $\cos \delta'$  9.93169 |
| a.c. $\log r$  7.76548 | a.c. $\log r$  7.76548 |
| $\log(t_2 - t_1)$  1.20140 | $\log(t_2' - t_1')$  1.26717 |
| $\sin \gamma$  9.77395 | $\sin \gamma'$  9.83940 |
| $\cos \gamma$  9.90542 | $\cos \gamma'$  9.85912 |
| $\log r$  2.23452 | $\log r$  2.23452 |
| $d$  $138''.0$ | $d'$  $124''.1$ |

$$\delta' - \delta = d' + d = + 262''.1 = + 4'\ 22''.1$$

These approximate values of the apparent differences $\alpha' - \alpha$ and $\delta' - \delta$ correspond to the Ann Arbor sidereal time 1888 Sept. 8, $19^h\ 19^m\ 22^s$.

# APPENDICES

### A. Hints on Computing

The numerical calculations required in the problems of practical astronomy are generally a source of discouragement to the beginner, even though he is a skilful mathematician. Practice in making extensive series of computations, however, very soon suggests to him various devices for avoiding much of the labor. Every computer acquires methods peculiarly his own; yet the following hints will possibly be useful to many.

Only those logarithmic tables should be employed which contain the auxiliary tables of proportional parts on the margins of the pages, excepting possibly three-place and four-place tables. They enable the computer to make nearly all the interpolations mentally; and the use of any other tables, for any purpose whatever, cannot be recommended.

The following are recommended:
*Bruhns's* or *Vega's* seven-place tables.
*Bremiker's* six-place tables.
*Hussey's*, *Newcomb's* or *Becker's* five-place tables.
*Zech's* addition and subtraction logarithmic tables.\*
*Barlow's* tables of squares, square-roots, etc.
*Crelle's* multiplication and division tables.

If extensive computations are made with seven-place tables and the interpolations carried to hundredths of sec-

---

\* *Bremiker's* tables contain *Gauss's* addition and subtraction logarithmic tables to six places. *Hussey's* and *Becker's* contain *Zech's* to five places, and *Newcomb's* contain *Gauss's* to five places.

onds, the results are usually accurate within a tenth of a second. If six-place tables are used and the interpolations carried to tenths of seconds, the results are usually accurate within a second. If five-place tables are employed and the interpolations carried to seconds or to hundredths of minutes, the results are usually accurate within five seconds.

*First of all*, an outline of the *whole solution* should be prepared by writing in a vertical column the symbols of all the functions that will be used. Those should be placed adjacent to each other which are to be combined, as shown by the formulæ. If a number of similar solutions for different values of the variable or variables are to be made, a vertical column should be arranged for each on the right of the column of functions, which thus serves for all, and the computations in the several columns should be carried on *simultaneously*. If the solutions are made for equidistant values of the variable or variables, this method affords a valuable check on the accuracy of the results, since all the quantities which are in the same horizontal line should differ systematically from each other as we go from the first column to the last. By subtracting the result in each column from the corresponding result in the next column to the right, any error will be detected very quickly by the fact that the *differences* will not vary properly. This method is called the **method of differences.** It will not detect systematic errors: that is, errors affecting all the columns alike.

If the sine, cosine, tangent, etc., of the same angle are required, they should all be taken from the tables at one opening. *Avoid turning twice to the same angle in the same solution.* The interpolations can be checked by subtracting mentally the last two figures of the cosine from those of the sine and comparing the result with the last two places of the tangent; and similarly in other cases.

The tangent of an angle always varies more rapidly

than its sine or cosine, and for this reason the value of an angle should be taken from the tables by means of its tangent if great accuracy is required.

Many of the operations can be performed mentally, thereby saving much time. Thus, two numbers can be added or subtracted mentally *from left to right*, or a number multiplied or divided by two from left to right, and the result held in mind while we turn to the tables and take out the proper angle or function. This has been done very largely in the solutions of the examples in this book. Just how far the student should carry the method depends upon the individual. The beginner will find it perplexing and a fruitful source of error, but after some practice he can perform the operations quickly and accurately. It should be said that many experienced computers prefer to set down the results in the usual way.

If there are several factors in the numerator or denominator of an expression to be evaluated, do not add the logarithms in the numerator together, those of the denominator together, and take the difference; but form the arithmetical complement of each logarithm in the denominator *mentally* by subtracting it from 10, from left to right, and set down the result in the proper place. All the factors can then be combined by one addition.

When a constant quantity is to be used several times, it should be written on the margin of a slip of paper and held over the quantities with which it is to be combined.

If two quantities are to be combined which are separated by one or more lines, hold a pencil or slip of paper over the intervening quantities and the two can then be combined as conveniently as if they were adjacent.

If two quantities are given by their logarithms, and the logarithm of their sum or difference is required, it should be found by means of addition and subtraction logarithmic tables. The result will be obtained more quickly and accurately than by means of the ordinary tables.

Whenever the formulæ furnish checks on the accuracy of the solution, they should generally be applied. The experienced computer usually detects an error very quickly.

If the trigonometrical function or other logarithmic function is negative, write the subscript $_n$ after it.

*Do not use negative characteristics.* Increase them by 10 if they are naturally negative.

The example in § 4 will illustrate many of these methods. First write down the two columns of functions, as the outline of the solution of (12), (13), (14) and the check equation (9). The values of $\phi$, $z$ and $A$ are inserted. From the tables $\tan z$ and $\sin z$ are found and written opposite their symbols; and likewise $\cos A$, $\tan A$ and $\sin A$. The sum of $\tan z$ and $\cos A$ is $\tan M$. Add them mentally, enter the tables and take out $M$. Take out $\sin M$ at once. Subtract $M$ from $\phi$, mentally, find $\sec(\phi - M)$ and $\tan(\phi - M)$. The value of $\sec(\phi - M)$ is $10 - \cos(\phi - M)$. Add the three logarithms to find $\tan t$. Determine $t$ from the tables and take $\cos t$ and $\csc t$ out. The sum of $\tan(\phi - M)$ and $\cos t$ is $\tan \delta$. Add them mentally and take $\delta$ and $\sec \delta$ from the tables. The sum of the last four logarithms is $\log 1$. It should not differ more than one or two units of the last place from zero or ten.

The printed solution contains every figure that need be written down. But possibly the beginner should write down $\tan M$, $\phi - M$ and $\tan \delta$.

B. INTERPOLATION FORMULÆ

The Ephemeris tabulates the values of any required function corresponding to equidistant values of the time. If the value of the function for any intermediate date is desired, it is sometimes determined most conveniently by means of a general interpolation formula.

Let $T$ be the date in the Ephemeris nearest the instant for which the value of the function is required, and let $\omega$ be the tabular interval of time. Then the adjacent dates

APPENDICES 245

in the Ephemeris may be represented as in the first column of the following table, and the corresponding values of the function as in the second column. Subtracting each value

| Argument | Function | 1st Diff. | 2d Diff. | 3d Diff. | 4th Diff. | 5th Diff. | 6th Diff. |
|---|---|---|---|---|---|---|---|
| $T - 3\omega$ | $f(T - 3\omega)$ | | | | | | |
| | | $a_{III}$ | | | | | |
| $T - 2\omega$ | $f(T - 2\omega)$ | | $b_{II}$ | | | | |
| | | $a_{II}$ | | $c_{II}$ | | | |
| $T - \omega$ | $f(T - \omega)$ | | $b_I$ | | $d_I$ | | |
| | | $a_I$ | | $c_I$ | | $e_I$ | |
| $T$ | $f(T)$ | $[a]$ | $b$ | $[c]$ | $d$ | $[e]$ | $f$ |
| | | $a'$ | | $c'$ | | $e'$ | |
| $T + \omega$ | $f(T + \omega)$ | | $b'$ | | $d'$ | | |
| | | $a''$ | | $c''$ | | | |
| $T + 2\omega$ | $f(T + 2\omega)$ | | $b''$ | | | | |
| | | $a'''$ | | | | | |
| $T + 3\omega$ | $f(T + 3\omega)$ | | | | | | |

of the function from the next following value, we obtain the "1st differences" in column three. Subtracting each 1st difference from the next following, we obtain the 2d differences in column four; and so on, for the differences of higher orders. Lastly, the quantities $[a] = \frac{1}{2}(a_I + a')$, $[c] = \frac{1}{2}(c_I + c')$ and $[e] = \frac{1}{2}(e_I + e')$ are inserted in the same horizontal line as $T$. Let the instant for which the value of the function is wanted be represented by $T + t$. Let $n$ be the ratio of $t$ and the tabular interval $\omega$; i.e., $n = t/\omega$, or $t = n\omega$. The value of the function required is

$$f(T + t) = f(T) + n[a] + \frac{n^2}{2}b + \frac{n(n^2-1)}{2 \cdot 3}[c] + \frac{n^2(n^2-1)}{2 \cdot 3 \cdot 4}d$$
$$+ \frac{n(n^2-1)(n^2-4)}{2 \cdot 3 \cdot 4 \cdot 5}[e] + \cdots.$$

*Example.* Determine from the American Ephemeris, page 219, the apparent declination of *Mercury* at Greenwich mean time 1899 April $2^d 16^h 0^m$.

246   PRACTICAL ASTRONOMY

The epoch $T$ is April $3^d.0$, the tabular interval $\omega$ is $24^h$, and $t$ is $-8^h$. Therefore, $n = -\frac{1}{3}$. The functions and differences are as below:

| Argument | Function | a | b | c | d | e |
|---|---|---|---|---|---|---|
| March $31^d.0$ | $+13°\ 16'\ 34''.4$ | | | | | |
| | | $+5'\ 55''.2$ | | | | |
| April 1 .0 | 13 22 29 .6 | | $-3'\ 55''.2$ | | | |
| | | $+2\ \ 0\ .0$ | | $+0''.8$ | | |
| 2 .0 | 13 24 29 .6 | | $-3\ 54\ .4$ | | $+2''.4$ | |
| | | $-1\ 54\ .4$ | | $+3\ .2$ | | $+0''.3$ |
| 3 .0 | 13 22 35 .2 | $[-3\ 50\ .0]$ | $-3\ 51\ .2$ | $[+4\ .6]$ | $+2\ .7$ | $[+0\ .1]$ |
| | | $-5\ 45\ .6$ | | $+5\ .9$ | | $-0\ .1$ |
| 4 .0 | 13 16 49 .6 | | $-3\ 45\ .3$ | | $+2\ .6$ | |
| | | $-9\ 30\ .9$ | | $+8\ .5$ | | |
| 5 .0 | 13 7 18 .7 | | $-3\ 36\ .8$ | | | |
| | | $-13\ \ 7\ .7$ | | | | |
| 6 .0 | $+12\ 54\ 11\ .0$ | | | | | |

The quantities to be taken from this table are all included, with April $3^d.0$, between the two horizontal lines. Substituting these values and $n = -\frac{1}{3}$ in the above equation we obtain the following values of the individual terms, and their sum, respectively:

$$\begin{array}{rr} + & 13°\ 22'\ 35''.2 \\ + & 1\ 16\ .7 \\ - & 12\ .8 \\ + & 0\ .2 \\ - & 0\ .1 \\ & 0\ .0 \\ \hline f(T+t)\ \ + & 13°\ 23'\ 39''.2 \end{array}$$

It is often required to determine by interpolation the value of a function for a date midway between two tabular dates. The required value is determined as follows: From the arithmetical mean of the two values of the function corresponding to the two tabular dates, subtract one-eighth of the arithmetical mean of the second differences found on the same horizontal lines as the two dates, and add three one hundred and twenty-eighths of the arithmetical mean of the fourth differences found on the same horizontal lines.

*Example.* Determine the apparent declination of *Mercury* at Greenwich mean time 1899 April $3^d.5$.
From the above data

$$\tfrac{1}{2}(13° 22' 35''.2 + 13° 16' 49''.6) = + 13° 19' 42''.4$$
$$-\tfrac{1}{4} \cdot \tfrac{1}{2}(- \quad 3\ 51\ .2 - \quad 3\ 45\ .3) = + \qquad 28\ .5$$
$$+ \tfrac{3}{128} \cdot \tfrac{1}{2}(+ \quad 2\ .7 + \quad 2\ .6) = \qquad\qquad 0\ .0$$
$$\text{App. } \delta,\ 1899\ \text{April } 3^d.5 = + 13° 20' 10''.9$$

### C. Combination and Comparison of Observations

*Formulæ resulting from the Method of Least Squares*

1. Direct observations of a quantity: $n$ separate results, $m_1, m_2, \cdots m_n$ of equal weight.

Most probable value of quantity, $z = \dfrac{[m]}{n}.$ *

Residuals, $z - m_1 = v_1,\ z - m_2 = v_2, \cdots z - m_n = v_n.$

Probable error of $z$, $\qquad r_0 = \pm\, 0.6745 \sqrt{\dfrac{[vv]}{n(n-1)}}.$

Probable error of a single observation, $r = \pm\, 0.6745 \sqrt{\dfrac{[vv]}{n-1}}.$

2. Direct observations of a quantity: $n$ separate results, $m_1, m_2, \cdots m_n$ of unequal weights, $p_1, p_2, \cdots p_n.$

Most probable value of quantity, $\qquad z = \dfrac{[pm]}{[p]}.$

Probable error of $z$, $\qquad r_0 = \pm\, 0.6745 \sqrt{\dfrac{[pvv]}{[p](n-1)}}.$

Probable error of an obs'n of weight unity, $r = \pm\, 0.6745 \sqrt{\dfrac{[pvv]}{n-1}}.$

---

* The symbols [ ] signify the sum of all similar quantities. Thus
$$[m] \equiv m_1 + m_2 + \cdots + m_n.$$
$$[pvv] \equiv p_1 v_1^2 + p_2 v_2^2 + \cdots + p_n v_n^2.$$

Weight of $z$, $\qquad P = [p]$.

Relation of weights to probable errors, $\qquad p_1 : p_2 : \cdots :: \dfrac{1}{r_1^2} : \dfrac{1}{r_2^2} : \cdots$

3. If $Z = az_1 \pm bz_2 \pm \cdots kz_n$, and the probable errors and weights of $z_1, z_2, \cdots z_n$ are $r_1, r_2, \cdots r_n$ and $p_1, p_2, \cdots p_n$, then the probable error and weight of $Z$ are given by

$$r = \pm \sqrt{(ar_1)^2 + (br_2)^2 + \cdots (kr_n)^2},$$

$$\frac{1}{p} = \frac{a^2}{p_1} + \frac{b^2}{p_2} + \cdots \frac{k^2}{p_n}.$$

4. In general, if $Z = f(z_1, z_2, \cdots z_n)$, the probable error of $Z$ is

$$r = \pm \sqrt{\left(\frac{df}{dz_1}\right)^2 r_1^2 + \left(\frac{df}{dz_2}\right)^2 r_2^2 + \cdots + \left(\frac{df}{dz_n}\right)^2 r_n^2}.$$

5. Direct observations of a function of a quantity $z$: the separate results, $m_1, m_2, \cdots m_n$ of equal weight, and the form of the function, $az$. The observation equations are

$$a_1 z + m_1 = 0,$$
$$a_2 z + m_2 = 0,$$
$$\cdots \cdots \cdots$$
$$a_n z + m_n = 0.$$

The most probable value of $z$ and its probable error are

$$z = -\frac{[am]}{[aa]}, \qquad r = \pm\, 0.6745 \sqrt{\frac{[vv]}{[aa]\,(n-1)}}.$$

If the observations are of unequal weights, multiply the observation equations through by the square roots of their respective weights, and proceed as before.

6. Direct observations of a function of two quantities, $w$ and $z$: the separate results, $m_1, m_2, \ldots m_n$ of equal weights, and the form of the function, $aw + bz$. The observation equations are

$$a_1 w + b_1 z + m_1 = 0,$$
$$a_2 w + b_2 z + m_2 = 0,$$
$$\cdots \cdots \cdots \cdots$$
$$a_n w + b_n z + m_n = 0.$$

The normal equations are
$$[aa]w + [ab]z + [am] = 0,$$
$$[ab]w + [bb]z + [bm] = 0.$$
Let
$$[bb] - \frac{[ab]}{[aa]}[ab] = [bb.1], \quad [bm] - \frac{[ab]}{[aa]}[am] = [bm.1].$$
Then the most probable values of $w$ and $z$ are given by
$$z = -\frac{[bm.1]}{[bb.1]},$$
$$w = -\frac{[ab]}{[aa]}z - \frac{[am]}{[aa]}.$$
The weights of $w$ and $z$ are
$$p_z = [bb.1], \quad p_w = \frac{[bb.1]}{[bb]}[aa].$$
The probable error of a single observation (of weight unity) is
$$r = \pm\, 0.6745\sqrt{\frac{[vv]}{n-2}};$$
and the probable errors of $w$ and $z$ are
$$r_w = \frac{r}{\sqrt{p_w}}, \quad r_z = \frac{r}{\sqrt{p_z}}.$$

If the observations are of unequal weights, multiply the observation equations through by the square roots of their respective weights, and proceed as before.

7. Direct observations of a function of three quantities, $x$, $y$ and $z$: the separate results, $m_1, m_2, \ldots m_n$ of equal weight, and the form of the function, $ax + by + cz$. The observation equations are
$$a_1 x + b_1 y + c_1 z + m_1 = 0,$$
$$a_2 x + b_2 y + c_2 z + m_2 = 0,$$
$$\cdots \cdots \cdots \cdots \cdots$$
$$a_n x + b_n y + c_n z + m_n = 0.$$
The normal equations are
$$[aa]x + [ab]y + [ac]z + [am] = 0,$$
$$[ab]x + [bb]y + [bc]z + [bm] = 0,$$
$$[ac]x + [bc]y + [cc]z + [cm] = 0.$$

Let

$$[bb] - \frac{[ab]}{[aa]}[ab] = [bb.1], \quad [bc] - \frac{[ab]}{[aa]}[ac] = [bc.1],$$

$$[bm] - \frac{[ab]}{[aa]}[am] = [bm.1],$$

$$[cc] - \frac{[ac]}{[aa]}[ac] = [cc.1], \quad [cm] - \frac{[ac]}{[aa]}[am] = [cm.1],$$

$$[cc.1] - \frac{[bc.1]}{[bb.1]}[bc.1] = [cc.2], \quad [cm.1] - \frac{[bc.1]}{[bb.1]}[bm.1] = [cm.2].$$

Then the most probable values of $x$, $y$ and $z$ are given by

$$z = -\frac{[cm.2]}{[cc.2]},$$

$$y = -\frac{[bc.1]}{[bb.1]}z - \frac{[bm.1]}{[bb.1]},$$

$$x = -\frac{[ab]}{[aa]}y - \frac{[ac]}{[aa]}z - \frac{[am]}{[aa]}.$$

The weights of $x$, $y$ and $z$ are given by

$$p_z = [cc.2],$$

$$p_y = \frac{[cc.2]}{[cc.1]}[bb.1],$$

$$p_x = \frac{[cc.2]}{[cc.1]_a} \cdot \frac{[bb.1]}{[bb]}[aa],$$

in which

$$[cc.1]_a = [cc] - \frac{[bc]}{[bb]}[bc].$$

The probable error of a single observation (of weight unity) is

$$r = \pm 0.6745\sqrt{\frac{[vv]}{n-3}},$$

and the probable errors of $x$, $y$ and $z$ are

$$r_x = \frac{r}{\sqrt{p_x}}, \quad r_y = \frac{r}{\sqrt{p_y}}, \quad r_z = \frac{r}{\sqrt{p_z}}.$$

If the observations are of unequal weights multiply the observation equations through by the square roots of their respective weights, and proceed as before.

APPENDICES 251

### D. Objects for the Telescope

Besides the moon, the planets and the Milky Way, the objects in the following list will be of interest to the student. Fuller descriptions of them, with many valuable hints on the use of the telescope, can be found in *Webb's Celestial Objects for Common Telescopes*, which is an excellent guide for the observer. Every student should provide himself with a good star atlas. *Klein's Star Atlas*, or *Heis's Atlas Cœlestis* is recommended.

| $a$, 1900.0 | $\delta$, 1900.0 | Object: description: remarks |
|---|---|---|
| $0^h$ $37^m.3$ | $+40°$ $43'$ | The Great Nebula in Andromeda. One of the most interesting in the sky, large, $2\frac{1}{2}°$ by $4°$, easily visible to the naked eye. A small companion nebula lies $22'$ south. |
| 0  53 .4 | + 81  20 | $U$ Cephei, variable, $7^m.1$ to $9^m.2$, period $2^d.5$. |
| 1  18 .9 | + 67  36 | $\psi$ Cassiopeiæ, triple, $A\ 4^m.5$, $B\ 9^m$, $C\ 10^m$. $AB = 30''$, $BC = 3''$. |
| 1  22 .6 | + 88  46 | $\alpha$ Ursæ Minoris or Polaris, the standard $2^m$ star; a $9^m$ companion at $s = 18''.5$. |
| 1  48 .0 | + 18  48 | $\gamma$ Arietis, double, $4^m.5$ and $5^m$, $p = 179°$, $s = 8''$. |
| 1  57 .7 | + 41  51 | $\gamma$ Andromedæ, double, $3^m.5$ and $5^m.5$, $p = 63°$, $s = 10''$. The $5^m.5$ is also double, but close and difficult even for the largest telescopes. |
| 2  12 .0 | + 56  41 | Cluster in Perseus. A magnificent object with a low power. Another fine cluster 3 minutes east. |
| 2  14 .3 | −  3  26 | $o$ Ceti, interesting variable, irregular, $1^m.7$ to $9^m.5$, period about $331^d$. |
| 3   1 .7 | + 40  34 | $\beta$ Persei (Algol), interesting variable, $2^m.3$ to $3^m.5$, period $2^d$ $20^h$ $48^m$ $55^s$. |
| 3  40 .2 | + 23  27 | Nebula in the Pleiades, very faint and difficult, Merope in its north extremity. |
| 4   7 .6 | + 50  59 | Cluster in Perseus, good with low power. |
| 4   9 .6 | − 13   0 | Planetary nebula in Eridanus, circular, $12^m$ star in center. |
| 4  30 .2 | + 16  19 | $\alpha$ Tauri (Aldebaran), $1^m$ star, red. |
| 5   9 .3 | + 45  54 | $\alpha$ Aurigæ (Capella), $1^m$ star. |
| 5   0 .7 | −  8  19 | $\beta$ Orionis (Rigel), double, $1^m$ and $9^m$, $s = 9''.5$. The $9^m$ is a close double, very difficult even with the largest instruments. |
| 5  28 .5 | + 21  57 | Nebula in Taurus large, faint, oblong. |

| $\alpha$, 1900.0 | $\delta$, 1900.0 | Object: description: remarks |
|---|---|---|
| $5^h$ $30^m$.4 | $- 5° 27'$ | The Great Nebula in Orion, one of the most interesting nebulæ in the sky, about 3° by 5° in size. Near its densest part is the multiple star $\theta$ Orionis, called the Trapezium. The spectrum of the nebula indicates a gaseous composition. |
| 5  35 .7 | $- 2$  0 | $\zeta$ Orionis, triple, $A\,3^m$, $B\,6^m.5$, $C\,10^m$, $AB = 2''.5$, $AC = 57''$. |
| 6   2 .7 | $+ 24$ 21 | Cluster in Gemini, fine field with low power. |
| 6  37 .4 | $+ 59$ 33 | 12 Lyncis, triple, $A\,6^m$, $B\,6^m.5$, $C\,7^m.5$, $AB = 1''.5$, $AC = 8''.6$. |
| 6  40 .7 | $- 16$ 34 | $\alpha$ Canis Majoris (Sirius), the brightest star in the sky. A close 10 mag. companion is now (1899) difficult in powerful telescopes. |
| 7  14 .1 | $+ 22$ 10 | $\delta$ Geminorum, double, one yellow $3^m.5$, the other red $8^m$, $p = 205°$, $s = 7''$. |
| 7  28 .2 | $+ 32$  6 | $\alpha$ Geminorum (Castor), fine double, $3^m$ and $3^m.5$, $p = 220°$, $s = 5''.7$. |
| 7  34 .1 | $+  5$ 29 | $\alpha$ Canis Minoris (Procyon) $1^m$, with $13^m$ companion discovered in 1896, difficult with large instruments. At discovery, $p = 320°$, $s = 4''.7$. |
| 8  34 .5 | $+ 20$ 17 | Cluster in Cancer (Præsepe), fine field with low power. |
| 8  45 .7 | $+ 12$ 10 | Cluster in Cancer, about 200 stars, $9^m$ to $15^m$. |
| 9  47 .2 | $+ 69$ 36 | Nebulæ in Ursa Major, two nebulæ, 30' apart, preceding one brighter with bright nucleus. |
| 10  14 .4 | $+ 20$ 21 | $\gamma$ Leonis, fine double, $2^m$ and $3^m.5$. In 1897, $p = 115°$, $s = 3''.8$. |
| 10  19 .9 | $- 17$ 39 | Planetary Nebula in Hydra, fairly bright. |
| 11  12 .5 | $+ 59$ 19 | Nebula in Ursa Major, small, bright, with nucleus. |
| 11  47 .7 | $+ 37$ 33 | Nebula in Ursa Major, bright, 3' to 4' in diameter. |
| 12   5 .0 | $+ 19$  6 | Cluster in Coma Berenices, globular, bright, well resolved in large telescopes. |
| 12  34 .8 | $- 11$  4 | Nebula in Virgo, elliptical, 30'' by 5', fine field with low power. |
| 12  36 .6 | $-  0$ 54 | $\gamma$ Virginis, double, $4^m$ and $4^m$. In 1898, $p = 330°$, $s = 6''$. |
| 13  19 .9 | $+ 55$ 27 | $\zeta$ Ursæ Majoris, fine double, $3^m$ and $5^m$, $s = 14''$. |
| 13  37 .5 | $+ 28$ 52 | Cluster in Canes Venatici, bright, globular, probably more than 1,000 stars. |
| 14  11 .1 | $+ 19$ 43 | $\alpha$ Boötis (Arcturus), $1^m$ star, yellow. |

APPENDICES 253

| $a$, 1900.0 | $\delta$, 1900.0 | Object: description: remarks |
|---|---|---|
| $14^h$ $40^m.6$ | $+ 27°$ $30'$ | $\epsilon$ *Boötis*, beautiful double, $3^m$ yellow and $7^m$ blue, $s = 3''$, $p = 328°$. |
| 15 14 .1 | + 32 1 | $U$ *Coronæ*, variable, $7^m.5$ to $8^m.9$, period $3^d$ $10^h$ $51^m$. |
| 16 23 .3 | − 26 13 | $a$ *Scorpii*, double, $1^m$ and $7^m$, $s = 3''$. |
| 16 37 .5 | + 31 47 | $\zeta$ *Herculis*, double, $3^m$ and $6^m$, $s = 0''.6$ in 1899, period about 35 years, now (1899) very difficult in large instruments. |
| 16 38 .1 | + 36 39 | The *Cluster in Hercules*, globular, one of the finest of its kind. [eter. |
| 16 40 .3 | + 23 59 | *Nebula in Hercules*, planetary, $8''$ in diam- |
| 17 10 .1 | + 14 30 | $a$ *Herculis*, variable, $3^m.1$ to $3^m.9$, irregular period; companion $5^m.5$ at $p = 110°$, $s = 4''.7$. |
| 17 11 .5 | + 1 19 | $U$ *Ophiuchi*, $6^m.0$ to $6^m.7$, period $20^h$ $8^m$. |
| 17 51 .1 | − 18 59 | *Cluster in Ophiuchus*, good field with low power. |
| 17 58 .6 | + 66 38 | *Nebula in Draco*, planetary, bright, diameter $35''$, very near pole of ecliptic, very interesting. |
| 18 7 .3 | + 6 50 | *Nebula in Ophiuchus*, planetary, bright, diameter $5''$. |
| 18 33 .6 | + 38 41 | $a$ *Lyræ* (*Vega*), brightest star in northern hemisphere. |
| 18 41 .0 | + 39 34 | $\epsilon$ *Lyræ*, a multiple star, $A$ $5^m$, $B$ $6^m.5$, $C$ $5^m$, $D$ $5^m.5$, $AB$ $3''$, $CD$ $2''.3$, $AC$ $207''$. Numerous small stars between $AB$ and $CD$. |
| 18 46 .4 | + 33 15 | $\beta$ *Lyræ*, variable, $3^m.4$ to $4^m.5$, period $12^d$ $21^h$ $47^m$. |
| 18 49 .8 | + 32 54 | *Ring Nebula in Lyra*, annular, gaseous, most interesting of its kind. |
| 19 26 .7 | + 27 45 | $\beta$ *Cygni*, fine double, $3^m$ yellow and $7^m$ blue, $p = 56°$, $s = 35''$. |
| 19 48 .5 | + 70 1 | $\epsilon$ *Draconis*, double, $5^m.5$ and $7^m.5$, $s = 3''$. |
| 19 55 .2 | + 22 27 | *Nebula in Vulpecula*, the "Dumb Bell Nebula," double, large. |
| 20 42 .0 | + 15 46 | $\gamma$ *Delphini*, double, $4^m$ and $6^m$, $s = 11''$. |
| 20 58 .7 | − 11 45 | *Nebula in Aquarius*, planetary, bright, very interesting in a large telescope. |
| 21 2 .4 | + 38 15 | 61 *Cygni*, double, $5^m.5$ and $6^m$, $s = 21''$, one of the nearest stars to us. |
| 21 8 .2 | + 68 5 | $T$ *Cephei*, variable, $5^m.6$ to $9^m.0$, period $383^d$. |
| 21 28 .2 | − 1 16 | *Cluster in Aquarius*, large, globular. |
| 22 23 .7 | − 0 32 | $\zeta$ *Aquarii*, double, $4^m$ and $4^m.5$, $s = 3''.3$ in 1897. |
| 23 21 .1 | + 41 59 | *Nebula in Andromeda*, planetary, small, very bright, round. |

## TABLE I. PULKOWA REFRACTION TABLES

| App't z | log μ | λ | App't z | log μ | λ | App't z | log μ | λ | A |
|---|---|---|---|---|---|---|---|---|---|
| ° ′ | | | ° ′ | | | ° ′ | | | |
| 0 0 | 1.76032 | | 71 0 | 1.75614 | 1.0115 | 77 0 | 1.75131 | 1.0253 | 1.0029 |
| 5 0 | 1.76032 | | 10 | 1.75606 | 1.0118 | 10 | 1.75107 | 1.0259 | 1.0029 |
| 10 0 | 1.76030 | | 20 | 1.75598 | 1.0120 | 20 | 1.75083 | 1.0264 | 1.0030 |
| 15 0 | 1.76028 | | 30 | 1.75590 | 1.0123 | 30 | 1.75058 | 1.0271 | 1.0030 |
| 20 0 | 1.76025 | | 40 | 1.75582 | 1.0125 | 40 | 1.75032 | 1.0278 | 1.0031 |
| 25 0 | 1.76021 | | 50 | 1.75573 | 1.0128 | 50 | 1.75005 | 1.0285 | 1.0032 |
| 30 0 | 1.76015 | | 72 0 | 1.75564 | 1.0130 | 78 0 | 1.74976 | 1.0293 | 1.0033 |
| 35 0 | 1.76006 | | 10 | 1.75555 | 1.0133 | 10 | 1.74947 | 1.0300 | 1.0033 |
| 40 0 | 1.75995 | | 20 | 1.75546 | 1.0136 | 20 | 1.74917 | 1.0309 | 1.0034 |
| 45 0 | 1.75980 | 1.0018 | 30 | 1.75536 | 1.0138 | 30 | 1.74886 | 1.0318 | 1.0035 |
| 50 0 | 1.75960 | 1.0022 | 40 | 1.75526 | 1.0141 | 40 | 1.74853 | 1.0327 | 1.0036 |
| 51 0 | 1.75955 | 1.0024 | 50 | 1.75516 | 1.0144 | 50 | 1.74819 | 1.0335 | 1.0037 |
| 52 0 | 1.75949 | 1.0025 | 73 0 | 1.75506 | 1.0147 | 79 0 | 1.74783 | 1.0344 | 1.0038 |
| 53 0 | 1.75943 | 1.0026 | 10 | 1.75496 | 1.0150 | 10 | 1.74746 | 1.0354 | 1.0039 |
| 54 0 | 1.75936 | 1.0027 | 20 | 1.75485 | 1.0153 | 20 | 1.74707 | 1.0364 | 1.0040 |
| 55 0 | 1.75928 | 1.0029 | 30 | 1.75474 | 1.0157 | 30 | 1.74665 | 1.0374 | 1.0041 |
| 56 0 | 1.75920 | 1.0032 | 40 | 1.75462 | 1.0160 | 40 | 1.74623 | 1.0385 | 1.0042 |
| 57 0 | 1.75912 | 1.0035 | 50 | 1.75450 | 1.0163 | 50 | 1.74579 | 1.0397 | 1.0043 |
| 58 0 | 1.75902 | 1.0038 | 74 0 | 1.75438 | 1.0166 | 80 0 | 1.74533 | 1.0409 | 1.0044 |
| 59 0 | 1.75892 | 1.0041 | 10 | 1.75425 | 1.0170 | 10 | 1.74484 | 1.0421 | 1.0045 |
| 60 0 | 1.75881 | 1.0044 | 20 | 1.75412 | 1.0173 | 20 | 1.74433 | 1.0433 | 1.0046 |
| 61 0 | 1.75868 | 1.0047 | 30 | 1.75398 | 1.0177 | 30 | 1.74380 | 1.0447 | 1.0048 |
| 62 0 | 1.75853 | 1.0051 | 40 | 1.75384 | 1.0181 | 40 | 1.74325 | 1.0461 | 1.0049 |
| 63 0 | 1.75837 | 1.0055 | 50 | 1.75369 | 1.0185 | 50 | 1.74266 | 1.0475 | 1.0050 |
| 64 0 | 1.75820 | 1.0059 | 75 0 | 1.75354 | 1.0188 | 81 0 | 1.74204 | 1.0491 | 1.0052 |
| 65 0 | 1.75801 | 1.0064 | 10 | 1.75338 | 1.0191 | 10 | 1.74139 | 1.0508 | 1.0053 |
| 66 0 | 1.75780 | 1.0070 | 20 | 1.75322 | 1.0195 | 20 | 1.74071 | 1.0525 | 1.0055 |
| 67 0 | 1.75755 | 1.0077 | 30 | 1.75306 | 1.0200 | 30 | 1.73999 | 1.0542 | 1.0057 |
| 68 0 | 1.75727 | 1.0085 | 40 | 1.75289 | 1.0205 | 40 | 1.73924 | 1.0561 | 1.0059 |
| 69 0 | 1.75694 | 1.0093 | 50 | 1.75271 | 1.0211 | 50 | 1.73844 | 1.0580 | 1.0061 |
| 70 0 | 1.75657 | 1.0103 | 76 0 | 1.75253 | 1.0216 | 82 0 | 1.73760 | 1.0600 | 1.0063 |
| 10 | 1.75650 | 1.0105 | 10 | 1.75235 | 1.0223 | 10 | 1.73671 | 1.0622 | 1.0065 |
| 20 | 1.75643 | 1.0107 | 20 | 1.75216 | 1.0229 | 20 | 1.73577 | 1.0645 | 1.0068 |
| 30 | 1.75636 | 1.0109 | 30 | 1.75196 | 1.0235 | 30 | 1.73478 | 1.0669 | 1.0070 |
| 40 | 1.75629 | 1.0111 | 40 | 1.75175 | 1.0241 | 40 | 1.73373 | 1.0694 | 1.0073 |
| 50 | 1.75622 | 1.0113 | 50 | 1.75153 | 1.0246 | 50 | 1.73260 | 1.0720 | 1.0076 |
| 71 0 | 1.75614 | 1.0115 | 77 0 | 1.75131 | 1.0253 | 83 0 | 1.73143 | 1.0747 | 1.0078 |

### Supplement

| App't z | log μ tan z | λ | A | App't z | log μ tan z | λ | A |
|---|---|---|---|---|---|---|---|
| ° ′ | | | | ° ′ | | | |
| 82 30 | 2.61534 | 1.0669 | 1.0070 | 86 30 | 2.88535 | 1.1934 | 1.0203 |
| 83 0 | 2.64226 | 1.0747 | 1.0078 | 87 0 | 2.93113 | 1.2277 | 1.0241 |
| 83 30 | 2.67076 | 1.0839 | 1.0087 | 87 30 | 2.98087 | 1.2708 | 1.0294 |
| 84 0 | 2.70088 | 1.0949 | 1.0098 | 88 0 | 3.03519 | 1.3241 | 1.0357 |
| 84 30 | 2.73294 | 1.1080 | 1.0112 | 88 30 | 3.09458 | 1.3902 | 1.0437 |
| 85 0 | 2.76717 | 1.1235 | 1.0127 | 89 0 | 3.15004 | 1.4729 | 1.0541 |
| 85 30 | 2.80376 | 1.1424 | 1.0148 | 89 30 | 3.23206 | 1.5762 | 1.0680 |
| 86 0 | 2.84304 | 1.1652 | 1.0172 | 90 0 | 3.31186 | 1.7046 | 1.0859 |

APPENDICES 255

TABLE I. PULKOWA REFRACTION TABLES

B. *Factor depending on the Barometer*

| English inches | log B | English inches | log B | French metres | log B |
|---|---|---|---|---|---|
| 25.0 | − 0.07330 | 28.0 | − 0.02409 | 0.724 | − 0.01621 |
| 25.1 | − 0.07157 | 28.1 | − 0.02254 | 0.726 | − 0.01500 |
| 25.2 | − 0.06984 | 28.2 | − 0.02099 | 0.728 | − 0.01380 |
| 25.3 | − 0.06812 | 28.3 | − 0.01946 | 0.730 | − 0.01261 |
| 25.4 | − 0.06641 | 28.4 | − 0.01793 | 0.732 | − 0.01142 |
| 25.5 | − 0.06470 | 28.5 | − 0.01640 | 0.734 | − 0.01024 |
| 25.6 | − 0.06300 | 28.6 | − 0.01488 | 0.736 | − 0.00906 |
| 25.7 | − 0.06131 | 28.7 | − 0.01336 | 0.738 | − 0.00788 |
| 25.8 | − 0.05962 | 28.8 | − 0.01185 | 0.740 | − 0.00670 |
| 25.9 | − 0.05794 | 28.9 | − 0.01035 | 0.742 | − 0.00553 |
| 26.0 | − 0.05627 | 29.0 | − 0.00885 | 0.744 | − 0.00436 |
| 26.1 | − 0.05460 | 29.1 | − 0.00735 | 0.746 | − 0.00319 |
| 26.2 | − 0.05294 | 29.2 | − 0.00586 | 0.748 | − 0.00203 |
| 26.3 | − 0.05129 | 29.3 | − 0.00438 | 0.750 | − 0.00087 |
| 26.4 | − 0.04964 | 29.4 | − 0.00290 | 0.752 | + 0.00028 |
| 26.5 | − 0.04800 | 29.5 | − 0.00142 | 0.754 | + 0.00144 |
| 26.6 | − 0.04636 | 29.6 | + 0.00005 | 0.756 | + 0.00259 |
| 26.7 | − 0.04473 | 29.7 | 0.00151 | 0.758 | + 0.00374 |
| 26.8 | − 0.04311 | 29.8 | 0.00297 | 0.760 | + 0.00488 |
| 26.9 | − 0.04149 | 29.9 | 0.00443 | 0.762 | + 0.00602 |
| 27.0 | − 0.03988 | 30.0 | 0.00588 | 0.764 | + 0.00716 |
| 27.1 | − 0.03827 | 30.1 | 0.00732 | 0.766 | + 0.00830 |
| 27.2 | − 0.03667 | 30.2 | 0.00876 | 0.768 | + 0.00943 |
| 27.3 | − 0.03508 | 30.3 | 0.01020 | 0.770 | + 0.01056 |
| 27.4 | − 0.03349 | 30.4 | 0.01163 | 0.772 | + 0.01168 |
| 27.5 | − 0.03191 | 30.5 | 0.01306 | 0.774 | + 0.01281 |
| 27.6 | − 0.03033 | 30.6 | 0.01448 | 0.776 | + 0.01393 |
| 27.7 | − 0.02876 | 30.7 | 0.01589 | 0.778 | + 0.01505 |
| 27.8 | − 0.02720 | 30.8 | 0.01731 | 0.780 | + 0.01616 |
| 27.9 | − 0.02564 | 30.9 | 0.01871 | 0.782 | + 0.01727 |
| 28.0 | − 0.02409 | 31.0 | + 0.02012 | 0.784 | + 0.01837 |

T. *Factor depending on Attached Thermometer*

| Fahr. | log T | Cent. | log T' |
|---|---|---|---|
| − 20° | + 0.00201 | − 30° | + 0.00209 |
| − 10 | 0.00162 | − 25 | 0.00174 |
| 0 | 0.00123 | − 20 | 0.00139 |
| + 10 | 0.00085 | − 15 | 0.00104 |
| 20 | 0.00047 | − 10 | 0.00069 |
| 30 | + 0.00008 | − 5 | + 0.00035 |
| 40 | − 0.00030 | 0 | 0.00000 |
| 50 | − 0.00069 | + 5 | − 0.00035 |
| 60 | − 0.00108 | 10 | − 0.00069 |
| 70 | − 0.00146 | 15 | − 0.00104 |
| 80 | − 0.00184 | 20 | − 0.00138 |
| 90 | − 0.00222 | 25 | − 0.00173 |
| + 100 | − 0.00262 | + 30 | − 0.00207 |

## TABLE I. PULKOWA REFRACTION TABLES

γ. *Factor depending on External Thermometer*

| Fahr. | log γ | Fahr. | log γ | Cent. | log γ |
|---|---|---|---|---|---|
| − 22° | + 0.06560 | + 35° | + 0.01200 | − 30° | + 0.06560 |
| − 21 | 0.06461 | 36 | 0.01112 | − 29 | 0.06381 |
| − 20 | 0.06361 | 37 | 0.01023 | − 28 | 0.06202 |
| − 19 | 0.06262 | 38 | 0.00935 | − 27 | 0.06023 |
| − 18 | 0.06162 | 39 | 0.00848 | − 26 | 0.05846 |
| − 17 | 0.06063 | 40 | 0.00760 | − 25 | 0.05669 |
| − 16 | 0.05964 | 41 | 0.00672 | − 24 | 0.05493 |
| − 15 | 0.05866 | 42 | 0.00585 | − 23 | 0.05317 |
| − 14 | 0.05767 | 43 | 0.00498 | − 22 | 0.05142 |
| − 13 | 0.05669 | 44 | 0.00411 | − 21 | 0.04968 |
| − 12 | 0.05571 | 45 | 0.00324 | − 20 | 0.04795 |
| − 11 | 0.05473 | 46 | 0.00238 | − 19 | 0.04622 |
| − 10 | 0.05376 | 47 | 0.00151 | − 18 | 0.04451 |
| − 9 | 0.05279 | 48 | + 0.00064 | − 17 | 0.04279 |
| − 8 | 0.05182 | 49 | − 0.00022 | − 16 | 0.04108 |
| − 7 | 0.05085 | 50 | − 0.00107 | − 15 | 0.03938 |
| − 6 | 0.04988 | 51 | − 0.00193 | − 14 | 0.03769 |
| − 5 | 0.04891 | 52 | − 0.00279 | − 13 | 0.03601 |
| − 4 | 0.04795 | 53 | − 0.00364 | − 12 | 0.03433 |
| − 3 | 0.04699 | 54 | − 0.00449 | − 11 | 0.03265 |
| − 2 | 0.04603 | 55 | − 0.00535 | − 10 | 0.03099 |
| − 1 | 0.04508 | 56 | − 0.00620 | − 9 | 0.02933 |
| 0 | 0.04413 | 57 | − 0.00704 | − 8 | 0.02767 |
| + 1 | 0.04318 | 58 | − 0.00789 | − 7 | 0.02602 |
| 2 | 0.04223 | 59 | − 0.00873 | − 6 | 0.02438 |
| 3 | 0.04128 | 60 | − 0.00957 | − 5 | 0.02274 |
| 4 | 0.04033 | 61 | − 0.01041 | − 4 | 0.02112 |
| 5 | 0.03938 | 62 | − 0.01125 | − 3 | 0.01950 |
| 6 | 0.03844 | 63 | − 0.01209 | − 2 | 0.01788 |
| 7 | 0.03750 | 64 | − 0.01293 | − 1 | 0.01627 |
| 8 | 0.03657 | 65 | − 0.01376 | 0 | 0.01466 |
| 9 | 0.03563 | 66 | − 0.01459 | + 1 | 0.01306 |
| 10 | 0.03470 | 67 | − 0.01543 | 2 | 0.01147 |
| 11 | 0.03377 | 68 | − 0.01626 | 3 | 0.00988 |
| 12 | 0.03284 | 69 | − 0.01709 | 4 | 0.00830 |
| 13 | 0.03191 | 70 | − 0.01792 | 5 | 0.00672 |
| 14 | 0.03099 | 71 | − 0.01874 | 6 | 0.00515 |
| 15 | 0.03007 | 72 | − 0.01956 | 7 | 0.00359 |
| 16 | 0.02915 | 73 | − 0.01838 | 8 | 0.00203 |
| 17 | 0.02822 | 74 | − 0.01920 | 9 | + 0.00047 |
| 18 | 0.02730 | 75 | − 0.02202 | 10 | − 0.00107 |
| 19 | 0.02639 | 76 | − 0.02284 | 11 | − 0.00261 |
| 20 | 0.02548 | 77 | − 0.02366 | 12 | − 0.00415 |
| 21 | 0.02456 | 78 | − 0.02447 | 13 | − 0.00569 |
| 22 | 0.02364 | 79 | − 0.02528 | 14 | − 0.00721 |
| 23 | 0.02273 | 80 | − 0.02609 | 15 | − 0.00873 |
| 24 | 0.02183 | 81 | − 0.02690 | 16 | − 0.01025 |
| 25 | 0.02094 | 82 | − 0.02771 | 17 | − 0.01176 |
| 26 | 0.02004 | 83 | − 0.02851 | 18 | − 0.01326 |
| 27 | 0.01914 | 84 | − 0.02932 | 19 | − 0.01476 |
| 28 | 0.01824 | 85 | − 0.03012 | 20 | − 0.01626 |
| 29 | 0.01734 | 86 | − 0.03093 | 21 | − 0.01775 |
| 30 | 0.01645 | 87 | − 0.03173 | 22 | − 0.01923 |
| 31 | 0.01555 | 88 | − 0.03253 | 23 | − 0.02071 |
| 32 | 0.01466 | 89 | − 0.03333 | 24 | − 0.02219 |
| 33 | 0.01377 | 90 | − 0.03413 | 25 | − 0.02366 |
| 34 | 0.01288 | 91 | − 0.03492 | 30 | − 0.03093 |
| + 35 | 0.01200 | 92 | − 0.03572 | + 35 | − 0.03810 |

APPENDICES 257

TABLE I. PULKOWA REFRACTION TABLES

| App't z | log σ | App't z | log σ | Date | i |
|---|---|---|---|---|---|
| ° ′ | | ° ′ | | | |
| 80 0 | 0.00019 | 85 0 | 0.00146 | Jan. 15 | + 0.34 |
| 80 30 | 0.00022 | 85 30 | 0.00185 | Feb. 15 | + 0.27 |
| 81 0 | 0.00025 | 86 0 | 0.00241 | Mar. 15 | + 0.05 |
| 81 30 | 0.00029 | 86 30 | 0.00320 | April 15 | − 0.08 |
| 82 0 | 0.00035 | 87 0 | 0.00421 | May 15 | − 0.20 |
| 82 30 | 0.00045 | 87 30 | 0.00561 | June 15 | − 0.26 |
| 83 0 | 0.00057 | 88 0 | 0.00749 | July 15 | − 0.33 |
| 83 30 | 0.00073 | 88 30 | 0.01006 | Aug. 15 | − 0.30 |
| 84 0 | 0.00091 | 89 0 | 0.01352 | Sept. 15 | − 0.19 |
| 84 30 | 0.00116 | 89 30 | 0.01813 | Oct. 15 | + 0.16 |
| 85 0 | 0.00146 | 90 0 | 0.02424 | Nov. 15 | + 0.33 |
| | | | | Dec. 15 | + 0.37 |

$$\log r = \log \mu + \log \tan z + A (\log B + \log T) + \lambda \log \gamma + i \log \sigma$$

TABLE II. PULKOWA MEAN REFRACTIONS
*Barom. 29.5 inches, Att. Therm. 50° F., Ext. Therm. 50° F.*

| App't z | Mean refr'n | App't z | Mean refr'n | App't z | Mean refr'n | App't z | Mean refr'n |
|---|---|---|---|---|---|---|---|
| ° ′ | ′ ″ | ° ′ | ′ ″ | ° ′ | ′ ″ | ° ′ | ′ ″ |
| 0 0 | 0 0.0 | 58 0 | 1 31.2 | 73 0 | 3 4.7 | 80 40 | 5 34 |
| 5 0 | 0 5.0 | 59 0 | 1 34.8 | 73 20 | 3 8.5 | 81 0 | 5 46 |
| 10 0 | 0 10.1 | 60 0 | 1 38.7 | 73 40 | 3 12.5 | 81 20 | 5 58 |
| 15 0 | 0 15.3 | 61 0 | 1 42.8 | 74 0 | 3 16.6 | 81 40 | 6 12 |
| 20 0 | 0 20.8 | 62 0 | 1 47.1 | 74 20 | 3 20.9 | 82 0 | 6 26 |
| 25 0 | 0 26.7 | 63 0 | 1 51.7 | 74 40 | 3 25.4 | 82 20 | 6 41 |
| 30 0 | 0 33.0 | 64 0 | 1 56.6 | 75 0 | 3 30.0 | 82 40 | 6 58 |
| 32 0 | 0 35.7 | 65 0 | 2 1.9 | 75 20 | 3 34.8 | 83 0 | 7 15 |
| 34 0 | 0 38.5 | 65 30 | 2 4.7 | 75 40 | 3 39.9 | 83 20 | 7 35 |
| 36 0 | 0 41.5 | 66 0 | 2 7.6 | 76 0 | 3 45.2 | 83 40 | 7 56 |
| 38 0 | 0 44.6 | 66 30 | 2 10.6 | 76 20 | 3 50.7 | 84 0 | 8 19 |
| 40 0 | 0 47.9 | 67 0 | 2 13.8 | 76 40 | 3 56.5 | 84 20 | 8 43 |
| 42 0 | 0 51.4 | 67 30 | 2 17.1 | 77 0 | 4 2.5 | 84 40 | 9 10 |
| 44 0 | 0 55.1 | 68 0 | 2 20.5 | 77 20 | 4 8.8 | 85 0 | 9 40 |
| 46 0 | 0 59.1 | 68 30 | 2 24.1 | 77 40 | 4 15.5 | 85 30 | 10 32 |
| 48 0 | 1 3.4 | 69 0 | 2 27.8 | 78 0 | 4 22.5 | 86 0 | 11 31 |
| 50 0 | 1 8.0 | 69 30 | 2 31.7 | 78 20 | 4 29.8 | 86 30 | 12 42 |
| 52 0 | 1 13.0 | 70 0 | 2 35.7 | 78 40 | 4 37.6 | 87 0 | 14 7 |
| 53 0 | 1 15.7 | 70 30 | 2 39.9 | 79 0 | 4 45.7 | 87 30 | 15 49 |
| 54 0 | 1 18.5 | 71 0 | 2 44.4 | 79 20 | 4 54.4 | 88 0 | 17 55 |
| 55 0 | 1 21.4 | 71 30 | 2 49.1 | 79 40 | 5 3.5 | 88 30 | 20 33 |
| 56 0 | 1 24.5 | 72 0 | 2 54.0 | 80 0 | 5 13.1 | 89 0 | 23 53 |
| 57 0 | 1 27.8 | 72 30 | 2 59.2 | 80 20 | 5 23.4 | 89 30 | 28 11 |
| 58 0 | 1 31.2 | 73 0 | 3 4.7 | 80 40 | 5 34.3 | 90 0 | 33 51 |

# PRACTICAL ASTRONOMY

TABLE III. $\quad m = \dfrac{2\sin^2 \tfrac{1}{2} t}{\sin 1''}$, or $m = \dfrac{2\sin^2 \tfrac{1}{2}(t_0 - t)}{\sin 1''}$.

| $t$, or $t_0 - t$ | $m$ | $t$, or $t_0 - t$ | $m$ | $t$, or $t_0 - t$ | $m$ | $t$, or $t_0 - t$ | $m$ | $t$, or $t_0 - t$ | $m$ |
|---|---|---|---|---|---|---|---|---|---|
| m s | " | m s | " | m s | " | m s | " | m s | " |
| 0  0 | 0.00 | 4  0 | 31.42 | 8  0 | 125.65 | 12  0 | 282.68 | 16  0 | 502.5 |
| 0  5 | 0.01 | 4  5 | 32.74 | 8  5 | 128.28 | 12  5 | 286.62 | 16  5 | 507.7 |
| 0 10 | 0.05 | 4 10 | 34.09 | 8 10 | 130.94 | 12 10 | 290.58 | 16 10 | 513.0 |
| 0 15 | 0.12 | 4 15 | 35.46 | 8 15 | 133.63 | 12 15 | 294.58 | 16 15 | 518.3 |
| 0 20 | 0.22 | 4 20 | 36.87 | 8 20 | 136.34 | 12 20 | 298.60 | 16 20 | 523.6 |
| 0 25 | 0.34 | 4 25 | 38.30 | 8 25 | 139.08 | 12 25 | 302.64 | 16 25 | 529.0 |
| 0 30 | 0.49 | 4 30 | 39.76 | 8 30 | 141.85 | 12 30 | 306.72 | 16 30 | 534.3 |
| 0 35 | 0.67 | 4 35 | 41.25 | 8 35 | 144.64 | 12 35 | 310.82 | 16 35 | 539.7 |
| 0 40 | 0.87 | 4 40 | 42.76 | 8 40 | 147.46 | 12 40 | 314.95 | 16 40 | 545.2 |
| 0 45 | 1.10 | 4 45 | 44.30 | 8 45 | 150.31 | 12 45 | 319.10 | 16 45 | 550.6 |
| 0 50 | 1.36 | 4 50 | 45.87 | 8 50 | 153.19 | 12 50 | 323.29 | 16 50 | 556.1 |
| 0 55 | 1.65 | 4 55 | 47.46 | 8 55 | 156.09 | 12 55 | 327.50 | 16 55 | 561.6 |
| 1  0 | 1.96 | 5  0 | 49.09 | 9  0 | 159.02 | 13  0 | 331.74 | 17  0 | 567.2 |
| 1  5 | 2.31 | 5  5 | 50.73 | 9  5 | 161.98 | 13  5 | 336.00 | 17  5 | 572.8 |
| 1 10 | 2.67 | 5 10 | 52.41 | 9 10 | 164.97 | 13 10 | 340.30 | 17 10 | 578.4 |
| 1 15 | 3.07 | 5 15 | 54.11 | 9 15 | 167.97 | 13 15 | 344.62 | 17 15 | 584.0 |
| 1 20 | 3.49 | 5 20 | 55.84 | 9 20 | 171.02 | 13 20 | 348.97 | 17 20 | 589.6 |
| 1 25 | 3.94 | 5 25 | 57.60 | 9 25 | 174.08 | 13 25 | 353.34 | 17 25 | 595.3 |
| 1 30 | 4.42 | 5 30 | 59.40 | 9 30 | 177.18 | 13 30 | 357.74 | 17 30 | 601.0 |
| 1 35 | 4.92 | 5 35 | 61.20 | 9 35 | 180.30 | 13 35 | 362.17 | 17 35 | 606.8 |
| 1 40 | 5.45 | 5 40 | 63.05 | 9 40 | 183.46 | 13 40 | 366.64 | 17 40 | 612.5 |
| 1 45 | 6.01 | 5 45 | 64.91 | 9 45 | 186.63 | 13 45 | 371.11 | 17 45 | 618.3 |
| 1 50 | 6.60 | 5 50 | 66.81 | 9 50 | 189.83 | 13 50 | 375.12 | 17 50 | 624.1 |
| 1 55 | 7.21 | 5 55 | 68.73 | 9 55 | 193.06 | 13 55 | 380.17 | 17 55 | 630.0 |
| 2  0 | 7.85 | 6  0 | 70.68 | 10  0 | 196.32 | 14  0 | 384.74 | 18  0 | 635.9 |
| 2  5 | 8.52 | 6  5 | 72.66 | 10  5 | 199.60 | 14  5 | 389.32 | 18  5 | 641.7 |
| 2 10 | 9.22 | 6 10 | 74.66 | 10 10 | 202.92 | 14 10 | 393.94 | 18 10 | 647.7 |
| 2 15 | 9.94 | 6 15 | 76.69 | 10 15 | 206.26 | 14 15 | 398.58 | 18 15 | 653.6 |
| 2 20 | 10.69 | 6 20 | 78.75 | 10 20 | 209.62 | 14 20 | 403.26 | 18 20 | 659.6 |
| 2 25 | 11.47 | 6 25 | 80.84 | 10 25 | 213.02 | 14 25 | 407.96 | 18 25 | 665.6 |
| 2 30 | 12.27 | 6 30 | 82.95 | 10 30 | 216.44 | 14 30 | 412.68 | 18 30 | 671.6 |
| 2 35 | 13.10 | 6 35 | 85.09 | 10 35 | 219.88 | 14 35 | 417.44 | 18 35 | 677.7 |
| 2 40 | 13.96 | 6 40 | 87.26 | 10 40 | 223.36 | 14 40 | 422.23 | 18 40 | 683.8 |
| 2 45 | 14.85 | 6 45 | 89.45 | 10 45 | 226.86 | 14 45 | 427.04 | 18 45 | 689.9 |
| 2 50 | 15.76 | 6 50 | 91.68 | 10 50 | 230.39 | 14 50 | 431.87 | 18 50 | 696.0 |
| 2 55 | 16.70 | 6 55 | 93.92 | 10 55 | 233.95 | 14 55 | 436.73 | 18 55 | 702.2 |
| 3  0 | 17.67 | 7  0 | 96.20 | 11  0 | 237.54 | 15  0 | 441.63 | 19  0 | 708.4 |
| 3  5 | 18.67 | 7  5 | 98.50 | 11  5 | 241.14 | 15  5 | 446.55 | 19  5 | 714.6 |
| 3 10 | 19.69 | 7 10 | 100.84 | 11 10 | 244.79 | 15 10 | 451.50 | 19 10 | 720.9 |
| 3 15 | 20.74 | 7 15 | 103.20 | 11 15 | 248.45 | 15 15 | 456.47 | 19 15 | 727.2 |
| 3 20 | 21.82 | 7 20 | 105.58 | 11 20 | 252.15 | 15 20 | 461.47 | 19 20 | 733.5 |
| 3 25 | 22.92 | 7 25 | 107.99 | 11 25 | 255.87 | 15 25 | 466.50 | 19 25 | 739.8 |
| 3 30 | 24.05 | 7 30 | 110.44 | 11 30 | 259.62 | 15 30 | 471.55 | 19 30 | 746.2 |
| 3 35 | 25.21 | 7 35 | 112.90 | 11 35 | 263.39 | 15 35 | 476.64 | 19 35 | 752.6 |
| 3 40 | 26.40 | 7 40 | 115.40 | 11 40 | 267.20 | 15 40 | 481.74 | 19 40 | 759.0 |
| 3 45 | 27.61 | 7 45 | 117.92 | 11 45 | 271.02 | 15 45 | 486.88 | 19 45 | 765.4 |
| 3 50 | 28.85 | 7 50 | 120.47 | 11 50 | 274.88 | 15 50 | 492.05 | 19 50 | 771.9 |
| 3 55 | 30.12 | 7 55 | 123.05 | 11 55 | 278.76 | 15 55 | 497.23 | 19 55 | 778.4 |
| 4  0 | 31.42 | 8  0 | 125.65 | 12  0 | 282.68 | 16  0 | 502.46 | 20  0 | 784.9 |

APPENDICES

TABLE III. $m = \dfrac{2\sin^2\frac{1}{2}t}{\sin 1''}$, or $m = \dfrac{2\sin^2\frac{1}{2}(t_0-t)}{\sin 1''}$. $\quad n = \dfrac{2\sin^4\frac{1}{2}t}{\sin 1''}$

| $t$, or $t_0 - t$ | | $m$ | $t$, or $t_0 - t$ | | $m$ | $t$ | | $n$ |
|---|---|---|---|---|---|---|---|---|
| m | s | " | m | s | " | m | s | " |
| 20 | 0 | 784.9 | 24 | 0 | 1129.9 | 0 | 0 | 0.00 |
| 20 | 5 | 791.4 | 24 | 5 | 1137.8 | 2 | 0 | 0.00 |
| 20 | 10 | 798.0 | 24 | 10 | 1145.6 | 4 | 0 | 0.00 |
| 20 | 15 | 804.6 | 24 | 15 | 1153.6 | 6 | 0 | 0.01 |
| 20 | 20 | 811.3 | 24 | 20 | 1161.5 | 8 | 0 | 0.04 |
| 20 | 25 | 817.9 | 24 | 25 | 1169.5 | 9 | 0 | 0.06 |
| 20 | 30 | 824.6 | 24 | 30 | 1177.5 | 10 | 0 | 0.09 |
| 20 | 35 | 831.2 | 24 | 35 | 1185.5 | 11 | 0 | 0.14 |
| 20 | 40 | 838.0 | 24 | 40 | 1193.5 | 12 | 0 | 0.19 |
| 20 | 45 | 844.7 | 24 | 45 | 1201.5 | 12 | 30 | 0.23 |
| 20 | 50 | 851.6 | 24 | 50 | 1209.6 | 13 | 0 | 0.26 |
| 20 | 55 | 858.4 | 24 | 55 | 1217.7 | 13 | 30 | 0.31 |
| 21 | 0 | 865.3 | 25 | 0 | 1225.9 | 14 | 0 | 0.36 |
| 21 | 5 | 872.1 | 25 | 5 | 1234.1 | 14 | 30 | 0.41 |
| 21 | 10 | 879.0 | 25 | 10 | 1242.3 | 15 | 0 | 0.47 |
| 21 | 15 | 886.0 | 25 | 15 | 1250.5 | 15 | 30 | 0.54 |
| 21 | 20 | 893.0 | 25 | 20 | 1258.8 | 16 | 0 | 0.61 |
| 21 | 25 | 900.0 | 25 | 25 | 1267.1 | 16 | 30 | 0.69 |
| 21 | 30 | 907.0 | 25 | 30 | 1275.4 | 17 | 0 | 0.78 |
| 21 | 35 | 914.0 | 25 | 35 | 1283.8 | 17 | 30 | 0.88 |
| 21 | 40 | 921.1 | 25 | 40 | 1292.2 | 18 | 0 | 0.98 |
| 21 | 45 | 928.2 | 25 | 45 | 1300.5 | 18 | 30 | 1.09 |
| 21 | 50 | 935.2 | 25 | 50 | 1309.0 | 19 | 0 | 1.22 |
| 21 | 55 | 942.3 | 25 | 55 | 1317.4 | 19 | 30 | 1.35 |
| 22 | 0 | 949.5 | 26 | 0 | 1325.9 | 20 | 0 | 1.49 |
| 22 | 5 | 956.7 | 26 | 5 | 1334.4 | 20 | 20 | 1.60 |
| 22 | 10 | 963.9 | 26 | 10 | 1342.9 | 20 | 40 | 1.70 |
| 22 | 15 | 971.2 | 26 | 15 | 1351.4 | 21 | 0 | 1.82 |
| 22 | 20 | 978.5 | 26 | 20 | 1360.1 | 21 | 20 | 1.93 |
| 22 | 25 | 985.8 | 26 | 25 | 1368.7 | 21 | 40 | 2.06 |
| 22 | 30 | 993.2 | 26 | 30 | 1377.3 | 22 | 0 | 2.19 |
| 22 | 35 | 1000.6 | 26 | 35 | 1385.9 | 22 | 20 | 2.32 |
| 22 | 40 | 1008.0 | 26 | 40 | 1394.7 | 22 | 40 | 2.46 |
| 22 | 45 | 1015.4 | 26 | 45 | 1403.4 | 23 | 0 | 2.61 |
| 22 | 50 | 1022.8 | 26 | 50 | 1412.2 | 23 | 20 | 2.77 |
| 22 | 55 | 1030.3 | 26 | 55 | 1420.9 | 23 | 40 | 2.93 |
| 23 | 0 | 1037.8 | 27 | 0 | 1429.7 | 24 | 0 | 3.10 |
| 23 | 5 | 1045.3 | 27 | 5 | 1438.5 | 24 | 20 | 3.27 |
| 23 | 10 | 1052.8 | 27 | 10 | 1447.4 | 24 | 40 | 3.45 |
| 23 | 15 | 1060.4 | 27 | 15 | 1456.3 | 25 | 0 | 3.64 |
| 23 | 20 | 1068.1 | 27 | 20 | 1465.2 | 25 | 20 | 3.84 |
| 23 | 25 | 1075.7 | 27 | 25 | 1474.1 | 25 | 40 | 4.05 |
| 23 | 30 | 1083.3 | 27 | 30 | 1483.1 | 26 | 0 | 4.26 |
| 23 | 35 | 1091.0 | 27 | 35 | 1492.1 | 26 | 20 | 4.48 |
| 23 | 40 | 1098.8 | 27 | 40 | 1501.1 | 26 | 40 | 4.72 |
| 23 | 45 | 1106.5 | 27 | 45 | 1510.2 | 27 | 0 | 4.96 |
| 23 | 50 | 1114.3 | 27 | 50 | 1519.2 | 27 | 20 | 5.20 |
| 23 | 55 | 1122.0 | 27 | 55 | 1528.3 | 27 | 40 | 5.46 |
| 24 | 0 | 1129.9 | 28 | 0 | 1537.5 | 28 | 0 | 5.73 |

# INDEX

(The Numbers refer to Pages)

Aberration: general, of a star in the direction of the observer's motion, 40; annual, defined, 41; annual, affecting a star's apparent place, 45, 61; annual, in right ascension and declination, 62, 64; diurnal, defined, 41; diurnal, in hour angle and declination, 41; diurnal, in azimuth, 43, 193, 200; diurnal, in azimuth and altitude, 43.

Altitude: defined, 3; measured at sea with a sextant, 36, 92; measured on land with a sextant, 93.

Angle: measured by vernier, 65; measured by reading microscope, 67, 179.

Apparent place: defined, 25, 45; reduction to, 61.

Axis of the celestial sphere, defined, 3.

Azimuth: defined, 3; constant of a transit instrument, 127,—determination of, 142; of a point, from observations of a star near elongation, 193; of a point, from observations of Polaris at any hour angle, 199; sometimes measured from north point, 199; of the polar axis of an equatorial telescope, 217.

Azimuth and altitude: as coördinates, 7; of a star, from given hour angle and declination, 10.

Barometer: factor in refraction formulæ, 34, 35, 255.

Berliner Astronomisches Jahrbuch, 2.

Bessel: his star numbers, 62; his star constants, 62.

Celestial sphere: defined, 2.

Chronograph: described, 86; illustration of, 87; used for comparing clocks and chronometers, 85.

Chronographic method of recording transit observations, 88.

Chronometer: correction, 83; rate, 83; care of, 85; transported for determining geographical longitude, 159; determination of correction from sextant observations, 101, — from transit observations, 127, 146, — from surveyor's transit observations, 203.

Circle: vertical, defined, 3; hour, defined, 3; latitude, defined, 5; primary, 7; secondary 7; graduated vertical, 175, 203; graduated horizontal, 191.

Circummeridian altitudes: of the sun, for determining geographical latitude, 112.

Clock: astronomical, 86; driving, 212.

Collimation: axis, 122; plane, 122; constant of a transit instrument, 127,—determination of, 137, 151.

Collimator: described, 138; determination of collimation constant by one, 138,—by two collimators, 141; determination of flexure by two collimators, 181.

Colure: defined, 4.

Connaissance des temps, 2.

Coördinates: spherical, 6; systems of, 7; transformation of, 8.

Day: sidereal, 15; true solar, 16; mean solar, 16; civil, 17; astronomical, 17.

Declination: defined, 4; fundamental determination of, 187; differential determination of, 187; axis of equatorial telescope, 212; circle of equatorial telescope, 212.

Dip of the horizon, 36.

Distance between two stars, 14.

Earth: form and dimensions of, 23; radius of, 25; reduction of observations to the center of, 23.
Eccentricity of a graduated circle, 69; of a sextant, determined, 97.
Ecliptic: defined, 4; obliquity of, 4; true, 45; mean, 45.
Elongation: of a star, 77, 193; reduction to, 194, 258.
Ephemeris: defined, 2; American, 2.
Equation of time: defined, 17.
Equator: celestial, defined, 3; reading of a circle, 187.
Equatorial telescope: described, 212; illustration of, 213; adjustments of, 214.
Equinoxes: defined, 4; precession of, 44.
Error of runs: defined, 68; determination of value of, 68, 190.
Eye and ear method of observing, 85.

Filar micrometer: described, 70; illustration of, 71; methods of using, 223, 229, 233.
Finder: of an equatorial telescope, 212; adjustment of, 215.

Geocentric place: defined, 25.
Graduated circle: read by vernier, 65; read by microscope, 67; eccentricity of, 69; graduation errors of, 183; flexure of, 184, 186.

Hints on computing, 241.
Horizon: defined, 2; dip of, 36; glass, 89; artificial, 93.
Hour angle: defined, 4; of a star, from given zenith distance and declination, 12; of a star, from given right ascension and sidereal time, 12.
Hour angle and declination: as coördinates, 7; of a star, from given azimuth and altitude, 8.
Hour circle: defined, 3; of an equatorial telescope, 212.

Index correction: of a sextant, defined, 96, — determined from observations of a star, 96, — from observations of the sun, 96.
Interpolation: elementary considerations on, 19; general formula for, 244; for a date midway between two tabular dates, 246.

Latitude: circles of, defined, 5; of a star, 5; geographical, 5; geocentric, 23; reduction to geocentric, 23; geographical, determined from meridian altitude of a star or the sun, 109, 209, — from an altitude of a star, 110, — from circummeridian altitudes, 112, — by Talcott's method, 167, — from meridian circle observations, 187, 188, 190; variation of, 174.
Least reading of a vernier, 66.
Least squares: application to determination of proper motion, 58, — to reduction of transit observations, 152, — to combination of latitude observations, 173; formulæ resulting from, 247.
Level: spirit, described, 79; general formulæ for, 80; adjustments of, 83; value of a division of, 81; striding, 123; zenith, 167; plate, 196, 211; constant of transit instrument, 127, — determination of, 134, 140; trier, 81, — illustration of, 81.
Longitude: of a star, defined, 5; geographical, defined, 5; geographical, determined by the method of lunar distances, 115, — by transportation of chronometers, 159, — by the electric telegraph, 160, — by the heliotrope, 163, — by moon culminations, 164.
Longitude and latitude: as coördinates, 7; of a star, from given right ascension and declination, 13.
Lunar distances: method of determining geographical longitude from, 115.

Meridian: defined, 3; direction of, determined from observations of a star, 196, 200.
Meridian circle: described, 175; illustration of, 176; flexure of, 181; determination of graduation errors of, 183; fundamental and differential determinations of declination, 187; determinations of latitude, 187, 188, 190.
Meridian mark, or mire, 143.
Micrometer: described, 70; filar, 70, — illustration of, 71; value of a revolution of, 72, — effect of temperature on the, 77; of a transit instrument,

123; of a zenith telescope, 125, 167; of a meridian circle, 179, 180; ring, 236.
Microscope: description of reading, 67, 177; methods of using, 67, 68, 179, 190; error of runs of, 68, 179, 190.
Mire, 143.

Nadir: defined, 2; determination of the level and collimation constants of a transit instrument by the method of the, 139, 140; reading of a meridian circle, 180.
Nautical Almanac, American Ephemeris and, 2.
Nautical Almanac (British), 2.
Noon: sidereal, 15; true solar, 16; mean solar, 16.
Nutation: defined, 44; affecting the true place of a star, 45; terms in reduction formulæ, 62, 64.

Objects for the telescope, 251.
Obliquity of the ecliptic: defined, 4; affected by attractions of the planets, 47.

Parallactic angle: defined, 11; of a star, from given azimuth and zenith distance, 11; of a star, from given hour angle and declination, 11.
Parallax: defined, 25; equatorial horizontal, 26; of a body in zenith distance, the earth being regarded as a sphere, 26; of a body in azimuth and zenith distance, the earth being regarded as a spheroid, 27; of a body in right ascension and declination, 31; factors, 32.
Personal equation: absolute, 157; relative, 157; machine, 159.
Plumb line: local deviations of, 23.
Polar axes: of an equatorial telescope, 212; adjustments of, 215, 217.
Polar distance: defined, 4.
Poles of the equator, 3.
Precession: luni-solar, 47; planetary, 47; general, 47; in right ascension and declination during any interval of time, 48; annual value of, in right ascension and declination, 50; secular variation of annual, 57.
Prime vertical: defined, 3.

Prismatic sextant, 91.
Prismatic transit instrument, 125.
Proper motion: of the stars, 45, 54; determination of annual, 54; reduced from one epoch to another, 55; determination of, by the method of least squares, 58.

Reduction to elongation: in the determination of azimuth, 194; tables for, 258.
Reduction to the meridian: of circum-meridian altitudes of a star or the sun, for latitude, 112, 258; for zenith telescope observations for latitude, 171; for meridian circle determinations of declination, 187.
Refraction: general laws of, 32; abnormal, in azimuth, 33; Pulkowa formula for, in zenith distance, 34; Pulkowa tables for computing, 254; Pulkowa mean, 257; Comstock's formula for, 35; differential formulæ for, in right ascension and declination, 36; differential, affecting zenith telescope observations for latitude, 169, 170; correction for, in meridian circle observations, 187; differential, affecting filar micrometer observations, 224, 229, 234.
Right ascension: defined, 5; of a star, from given hour angle and sidereal time, 12; of a star, from transit observations, 177; of the moon, from transit observations, 165.
Right ascension and declination: as coördinates, 7.
Ring micrometer: described, 236; determination of radius of, 236; method of observing with, 239.

Semidiameter: correction for, 38, 92, 93; apparent, of the moon, 38; contraction of, produced by refraction, 39.
Sequence and degree of corrections, 43.
Sextant: described, 89; illustration of, 90; general principles of, 90; prismatic, 91; methods of observing with, 92; adjustments of, 94; correction for index error of, 96; correction for eccentricity of, 97; determinations of time, 101; determinations of latitude, 109; observations of lunar dis-

tances for determining geographical longitude, 115.

Sidereal time: defined, 5; of a star, from given hour angle and right ascension, 12.

Solstices: defined, 4.

Spherometer: Harkness's, for investigating form of pivots, 137.

Stars: catalogues of, 46, 58, 59, 223; undetermined, 177; standard, 178; atlas of, 251.

Sun: mean, 16; right ascension of mean, 17; shades, to protect observer's eyes, 90, 207.

Surveyor's transit: determination of time by, 203, — of latitude by, 209, — of azimuth by, 211.

Talcott's method of determining latitude, 167.

Telescope: sextant, 89; zenith, 167; equatorial, 212; magnifying power of, 220; field of view of, 221.

Thermometer: attached, refraction factor for, 34, 255; external, refraction factor for, 34, 35, 256.

Time: sidereal, 15; true solar or apparent, 16; mean solar, 16; civil, 17; astronomical, 17; equation of, 17; conversion of, 18; determined from sextant observations, 101, — from star transits, 146, — from surveyor's transit observations, 203.

Transit instrument: described, 122; illustrations of, 124, 126; general formulæ for, 129; wire intervals of, 130, 132; reduction to middle wire of, 131; determination of level constant of, 134, 140, — of inequality of pivots of, 134, — of collimation constant of, 137, 151, — of azimuth constant of, 142; flexure of prismatic form of, 157.

Vernier: described, 65; illustration of, 66.

Year: tropical, 17; the fictitious, 46.

Zenith: defined, 2; level, 167; reading of a circle, 180.

Zenith distance: defined, 3; of a star, at greatest elongation, 78, 193.

Zenith telescope: Talcott's method of determining latitude by, 167.

www.ingramcontent.com/pod-product-compliance
Lightning Source LLC
Chambersburg PA
CBHW031947230426
43672CB00010B/2081